IMPORTANT WARNING
If the CD-ROM packaging has been opened,
The purchaser cannot return the book for a refund!
The CD-Rom is subject to this agreement!

LICENSING AND WARRANTY AGREEMENT

Notice to Users: Do not install or use the CD-ROM until you have read and agreed to this agreement. You will be bound by the terms of this agreement if you install or use the CD-ROM or otherwise signify acceptance of this agreement. If you do not agree to the terms contained in this agreement, do not install or use any portion of this CD-ROM.

License: The material in the CD-ROM (the "Software") is copyrighted and is protected by United States copyright laws and international treaty provisions. All rights are reserved to the respective copyright holders. No part of the Software may be reproduced, stored in a retrieval system, distributed (including but not limited to over the www/Internet), decompiled, reverse engineered, reconfigured, transmitted, or transcribed, in any form or by any means — electronic, mechanical, photocopying, recording, or otherwise — without the prior written permission of Brooks/Cole (the "Publisher"). Adopters of *Modern Geometry* by Dave Thomas may place the Software on the adopting school's network during the specific period of adoption for classroom purposes only in support of that text. The Software may not under any circumstances, be reproduced and/or downloaded for sale. For further permission and information, contact Brooks/Cole, 511 Forest Lodge Road, Pacific Grove, California 93950.

U.S. Government Restricted Rights: The enclosed Software and associated documentation are provided with RESTRICTED RIGHTS. Use, duplication, or disclosure by the Government is subject to restrictions as set forth in subdivision(c)(1)(ii) of the Rights in Technical Data and Computer Software clause at DFARS 252.277.7013 for DoD contracts, paragraphs(c)(1) and (2) of the Commercial Computer Software-Restricted Rights clause in the FAR (48 CFR 52.227-19) for civilian agencies, or in other comparable agency clauses. The proprietor of the enclosed software and associated documentation is Brooks/Cole, 511 Forest Lodge, Pacific Grove, CA 93950.

Limited Warranty: The warranty for the media on which the Software is provided is for ninety (90) days from the original purchase and valid only if the packaging for the Software was purchased unopened. If, during that time, you find defects in the workmanship or material, the Publisher will replace the defective media. The Publisher provides no other warranties, expressed or implied, including the implied warranties of merchantability or fitness for a particular purpose, and shall not be liable for any damages, including direct, special, indirect, incidental, consequential, or otherwise.

For Technical Support:
Voice: 1-800-423-0563
Fax: 1-859-647-5045
E-mail: support@kdc.com

D0169181

Modern Geometry

Modern Geometry

David A. Thomas
Ball State University

BROOKS/COLE

™

THOMSON LEARNING

Australia • Canada • Mexico • Singapore • Spain • United Kingdom • United States

BROOKS/COLE

THOMSON LEARNING

Publisher: *Bob Pirtle*
Marketing Team: *Karin Sandberg and Darcy Pool*
Editorial Assistant: *Molly Nance*
Project Editor: *Janet Hill*
Production Service: *WestWords, Inc./Patrick Burt*
Cover Design: *Roger Knox*

Interior Illustration: *Atherton Customs*
Print Buyer: *Vena Dyer*
Typesetting: *WestWords, Inc.*
Interior and Cover Printing/Binding:
 R. R. Donnelley & Sons,
 Crawfordsville

For more information, contact:

BROOKS/COLE

511 Forest Lodge Road

Pacific Grove, CA 93950 USA

www.brookscole.com

Printed in the United States of America

10   9   8   7   6   5   4   3   2   1

Library of Congress Cataloging-in-Publication Data

Thomas, David A. (David Allen)
    Modern geometry / David A. Thomas.
        p. cm.
    Includes bibliographical references and index.
    ISBN 0-534-36550-7
    1. Geometry. I. Title.

QA445.2 .T48 2001
516—dc21

To Cynthia, my wife and partner in adventure

Preface

The study of mathematics is one of mankind's most ancient and enduring intellectual pursuits. Although scraps of mathematical antiquity are found in the archaeological records of ancient Babylon and Egypt, mathematics first appeared as a distinct, formal discipline around 300 B.C., when Ptolomy I of Egypt opened the world's first great university at Alexandria, Egypt, and hired the Greek scholar Euclid to head the mathematics department. Working with other mathematicians and their students, Euclid created the world's first and most enduring axiomatic system of thought, Euclidean geometry. Euclid's most famous book, the *Elements,* might be the most translated, published, and studied book ever written, other than the *Bible.*

Since the time of Euclid, the study of geometry has been regarded as a foundation of western education and the preferred context in which to teach adolescents and young adults the purpose and value of logical thinking. In time, mathematicians discovered other, distinctly non-Euclidean geometries and applied them to the solution of new kinds of problems outside the scope of Euclidean geometry. Scientists and engineers have adopted these new geometries and used their power to unlock nature's secrets and adapt its resources to our needs. In the arts, painters, architects, and computer graphics designers have learned to create realistic 3-dimensional presentations on 2-dimensional surfaces such as television and theater screens using procedures that are fundamentally geometric yet non-Euclidean in nature. Unfortunately, these new geometries and their applications are, in most cases, overlooked in traditional secondary school mathematics curricula.

Modern Geometry was written to provide undergraduate and graduate level mathematics education students with an introduction to both Euclidean and non-Euclidean geometries appropriate to their needs as future junior and senior high school mathematics teachers. I believe that the need for this sort of training in many areas of mathematics is more critical today than at any time in recent history.

Across the country, a growing shortage of appropriately degreed and certified mathematics teachers threatens to undermine the very foundation of K–12 mathematics education—a well-prepared, professionally active mathematics teacher in every mathematics classroom. Unfortunately, many mathematics teachers lack both the mathematical content knowledge and the pedagogical content knowledge necessary to teach challenging mathematics to their students. In assessing the impact of this state of affairs on student achievement, the *Glenn Report 2000,* published by the U.S. Department of Education, states, "The evidence for the effect of better teaching is unequivocal. The most consistent and most powerful predictors of higher student achievement in mathematics and science are: (a) full certification of the teacher and (b) a college major in the field being taught. Conversely, the strongest predictors of lower student achievement are new teachers who are uncertified, or who hold less than a minor in their teaching field." The *Glenn Report* goes on to state that this problem is most serious in schools with the highest minority enrollments, where students have less than a 50% chance of getting a science or mathematics teacher holding degrees and teaching licenses in his/her assigned teaching areas.

Modern Geometry was developed to provide a new kind of tool with which to address one aspect of this complex problem. Specifically, *Modern Geometry* provides a systematic survey of Euclidean, hyperbolic, transformation, fractal, and projective geometries that is consistent with the recommendations of the National Council of Teachers of Mathematics (NCTM), the International Society for Technology in Education (ISTE), and other professional organizations active in the preparation and continuing professional development of K–12 mathematics teachers. In particular, *Modern Geometry* provides a technology-rich environment in which to investigate both Euclidean and non-Euclidean geometries, formulate and test conjectures, and solve problems.

Using powerful computer modeling tools such as the *Geometers Sketchpad* and WWW-based information resources such as the *MacTutor History of Mathematics Archives,* students investigate new concepts and procedures, review their historical context, and explore their value in a variety of mathematical and scientific contexts. Simultaneously, they learn essential definitions, axioms, and theorems and investigate their relationships. Through this process, students engage the geometric content both inductively and deductively and demonstrate their learning through a variety of exercises, constructions, investigations, problems, and proofs. In my experience, students taught using this approach respond enthusiastically and confidently to new challenges. They also develop an understanding of, and appreciation for, their discipline that provides an essential framework in which to address curriculum and instruction questions relevant to their future and/or present teaching assignments.

A variety of technologies have been employed to make the content of *Modern Geometry* more engaging and rewarding. For instance, conventional printed figures are static representations of particular configurations of objects in a relationship. Because textbook figures rarely address the range of possible configurations implicit in a given relationship, many students have difficulty seeing beyond any particular configuration to its underlying relationship. By making many of its figures available in the form of dynamic geometric models, *Modern Geometry* frees students to explore the underlying geometric relationships directly.

Students who purchase *Modern Geometry* will be able to explore many of the figures in the text by using JavaSketchpad. This tool for displaying dynamic geometric models in Java-enabled WWW browsers such as Internet Explorer and Netscape Navigator, is included on the free CD-ROM bound into every new copy of *Modern Geometry*. While the range of exploratory options provided by the JavaSketchpad is only a small subset of those provided by the Geometer's Sketchpad, it still reresents a significant instructional advantage over the use of print graphics alone. Those wishing to use the expanded options of Geometer's Sketchpad, may request a special bundle of the student edition with *Modern Geometry*.

Technology-based investigations and extensions to important concepts and procedures are available in more than half the sections of the book in the form of student investigations. The instructions, data files, and modeling tools for these investigations are available either on the *Modern Geometry* CD-ROM itself or as links on the *Modern Geometry* Website. This site provides a variety of HELP resources, corrections and additions to the text and CD-ROM materials, a discussion forum for *Modern Geometry* users, and samples of student work.

Because *Modern Geometry* is focused primarily on the needs of mathematics education majors, it may not be suitable for use in advanced geometry courses at the undergraduate and/or graduate levels. Enrollment in such courses is typically limited to mathematics majors (as opposed to mathematics education majors) having as their goal development as research mathematicians. This is neither the audience nor the goal for which *Modern Geometry* was developed. Having said this, I believe that *Modern Geometry* would serve well as the basis for an undergraduate mathematics elective and in a variety of professional development contexts for practicing mathematics teachers, including graduate mathematics education courses.

Written for use in the semester system, *Modern Geometry* may be adapted to use in the quarter system without losing the overall sense of coverage by omitting all or part of sections 4.5–4.6, 5.3–5.5, and 6.5–6.7.

I am sincerely interested in hearing from faculty and students regarding their experiences with *Modern Geometry,* corrections and suggestions, and examples of student work. For me, *Modern Geometry* always has been and always will be a work in progress. I look forward to hearing from you via the *Modern Geometry* Website: www.brookscole.com/thomasmoderngeometry.

Best wishes,

David A. Thomas

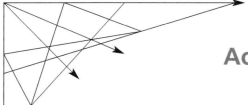

Acknowledgments

Modern Geometry could never have been written without the encouragement of my family, friends, colleagues, and students. They are the lens through which this text was first envisioned. Throughout the development of *Modern Geometry,* their feedback and advice have kept me true to my goal: Providing preservice and inservice secondary mathematics teachers with a technology-rich, historically grounded survey of ancient and modern geometries. While a comprehensive list of these supporters is neither practical nor possible, I want to single out the following individuals for special thanks: Cynthia Thomas, June Skillingberg, Howard Skillingberg, Maurice Burke, Brian Beaudrie, Robert Fixen, and Jeff Curry.

Thanks, too, to my reviewers for their useful comments on the manuscript: Thomas Butts, University of Texas–Dallas; Frederick Flener, Northeastern Illinois University; Stephen Pennel, University of Massachusetts, Lowell; Russell Rowlett, University of North Carolina–Chapel Hill; and Ron Solomon, Ohio State University.

The people of Brooks/Cole also have earned my sincere thanks and appreciation. In particular, I thank math publishers Gary Ostedt and Bob Pirtle and their dedicated staff. My first project for Brooks/Cole was *Active Geometry,* a technology supplement to James Smart's *Modern Geometry*, *Fifth Edition.* Shortly after the publication of *Active Geometry*, Gary Ostedt asked if I would undertake the writing of *Modern Geometry.* Since Gary's retirement, Bob Pirtle has continued to support the development of *Modern Geometry.*

The technology resources that accompany the *Modern Geometry* text include hundreds of computer models and data files created by myself and others. The range of interactive experiences facilitated by these resources is remarkable. Thanks to the generosity of the following individuals and corporations, students and teachers can explore Euclidean, transformation, hyperbolic, fractal, and projective geometry dynamically.

Individual/Corporation	Resource
Key Curriculum Press, Inc., Emeryville, CA	Geometers Sketchpad and JavaSketchpad
Edmund Robertson and John O'Connor, School of Mathematics and Statistics, St. Andrews University, Fife, Scotland	MacTutor History of Mathematics Archives

Individual/Corporation	Resource
United States Geological Survey, Reston, VA	dem3D and dlg32
Michael W. Frandsen, Helena, MT	Animate!
Roger Holmberg, Department of Computer and System Sciences, Stockholm University, Kista, Sweden	The Chaos Game
Paul Trunfio, Center for Polymer Studies, Boston University, Boston, MA	Fractal Coastline
Richard Horne, Microcomputer Topography, Visualization Software LLC	3dem60
Scion Corporation, Frederick, MD	Scion Image
Joel Castellanos, Albuquerque, NM	NonEuclid
Artifice, Inc., Eugene, OR	Design Workshop Lite
Geomantics Ltd., Lochearnhead, Scotland	Genesis II
Geometry Technologies, Inc., St. Paul, MN	Java Kali
Softronics, Inc.	MSW Logo

Contents

Chapter 1

Geometry Through the Ages

**URL
1.1.1**

The word geometry is derived from the Greek words "geo" meaning earth and "metry" meaning measure. While the ancient Greeks were the first to organize the study of geometric topics into a formal system of thought, mankind's interest in geometry predates the Greeks by many centuries. Many scholars believe that mathematics originated in the Orient (that is, in countries east of Greece) as an aid to business, agriculture, architecture, and engineering. For instance, clay tablets from the Sumerian (2100 B.C.) and Babylonian cultures (1600 B.C.) include a variety of mathematical tables for computing products, reciprocals, squares, square roots, and other mathematical functions useful in financial calculations. Examples of Babylonian geometric calculations include the areas of rectangles, right and isosceles triangles, trapezoids, and circles (computed as the square of the circumference divided by twelve). The Babylonians were also responsible for dividing the circumference of a circle into 360 equal parts. In addition to these accomplishments, the Babylonians also used the Phythagorean Theorem (long before Pythagoras), performed calculations involving ratio and proportion, and studied the relationships between the elements of various triangles. Ingenuity in devising mathematical tables and applying their mathematics to a wide range of problems is characteristic of Babylonian mathematics.

While not as mathematically inventive as the Babylonians, the ancient Egyptians used mathematics, particularly geometry, extensively. For instance, the Great Pyramid at Giza illustrates the precision they were capable of achieving. Covering approximately thirteen acres and containing over 2 million stone blocks, the sides of the base are said to be accurate to within one part in 14,000 and the right angles at the corners to within one part in 27,000. Egyptian mathematicians computed the area of the circle as the square of 8/9 the diameter. They also used the Pythagorean Theorem and computed the volumes and dihedral angles of pyramids and cylinders.

To the ancient Babylonians and Egyptians, the fundamental purpose of mathematics was to answer questions beginning with the word *how*. *How* can I compute the interest due on a loan? *How* can I compute the volume of a pyramid of given dimensions? *How* can I determine the square root of a given number? And so on. Questions such as these dominated ancient mathematics until approximately 1000 B.C. Around 1000 B.C., written language and monetary systems based on coins began revolutionizing intellectual and economic life in the Orient and the Middle East. Simultaneously, mathematical questions of a new kind began arising based on the word *why*. *Why* is the area of a triangle one-half the product of the base and the height? *Why* is the square of the hypotenuse of a right triangle equal to the sum of the squares of its sides? And so on. It may be that this revolution in mathematics appeared simultaneously with a rising tide of general intellectual and economic energy. Whatever the cause, mathematics changed forever. The prime movers in that revolution were the Greeks.

1.1   Greek Geometry Before Euclid

In this section, you will . . .

- **Be introduced to some of the ancient world's most famous geometers and philosophers.**
- **Investigate a few famous results of ancient mathematics.**

URL 1.1.2

In the history of Greek mathematics the following mathematicians played important roles in the development of new mathematics and the training of new mathematicians, conducted in schools founded and directed by the mathematicians themselves. A comprehensive list of these individuals and a timeline showing their relative lifetimes are available on-line at the MacTutor History of Mathematics Archive at St. Andrews University, Fife, Scotland. This outstanding WWW site is referenced throughout this book as a source of historical and contextual information.

Thales (640–546 B.C.)

URL 1.1.3

Thales of Miletus is widely viewed as the "father" of demonstrative mathematics. An engineer by trade, Thales took the practical geometric insights of the Egyptians, abstracted their mathematical features from the engineering contexts in which they were developed, asked *why* they were true, and used logical reasoning to demonstrate that truth. Proclus, the last major Greek philosopher (ca. fifth century A.D.), credits Thales with demonstrating the following theorems:

- A circle is bisected by any diameter.
- The base angles of an isosceles triangle are equal.
- The angles between two intersecting straight lines are equal.
- Two triangles are congruent if they have two angles and one side equal.
- An angle inscribed in a semicircle is a right angle.

Pythagoras (569–475 B.C.)

URL 1.1.4

Tradition holds that Pythagoras, born on the island of Samos, visited Thales as a youth and that this visit made a profound impression on Pythagoras. It is known that, years later, after traveling in both Europe and the Orient, Pythagoras settled in a Greek community in southern Italy and founded a "brotherhood" of a quasi-religious nature known as the Pythagoreans. This brotherhood was based on beliefs that

- At its deepest level, reality is mathematical in nature.
- Philosophy can be used for spiritual purification.
- The soul can rise to union with the divine.
- Certain symbols have a mystical significance.
- All brothers of the order should observe strict loyalty and secrecy.

Concerning the mathematical achievements of the Pythagoreans, it is believed that they were the first to demonstrate

- The sum of the angles of a triangle is equal to two right angles.
- For a right triangle, the square on the hypotenuse is equal to the sum of the squares on the other two sides.

In Figure 1.1.1, this is illustrated by the fact that the area of the square on the hypotenuse of $\triangle ABC$ is equal to the sum of the areas of the squares on the other two sides. Using modern notation, $c^2 = a^2 + b^2$.

CD
1.1.1

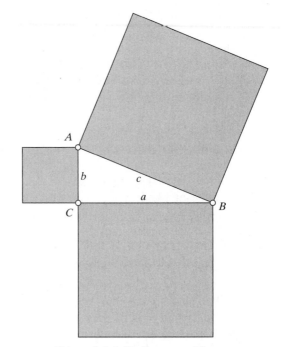

Figure 1.1.1 Pythagorean Theorem

- The existence of irrational numbers. If sides AC and BC in Figure 1.1.2 have length 1, the hypotenuse has length $\sqrt{2}$. Using elementary number theory, this length may be shown to be irrational (see exercises).

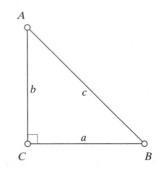

Figure 1.1.2 Isosceles Right Triangle

- Geometric solutions to quadratic equations. Figures 1.1.3–1.1.6 illustrate the approach taken by the Pythagoreans when investigating geometric approaches to what we regard today as algebraic relationships. Explaining these figures is left as an exercise.

CD 1.1.2

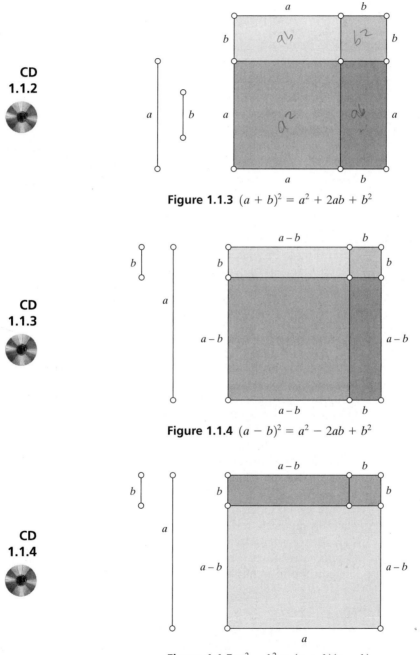

Figure 1.1.3 $(a + b)^2 = a^2 + 2ab + b^2$

CD 1.1.3

Figure 1.1.4 $(a - b)^2 = a^2 - 2ab + b^2$

CD 1.1.4

Figure 1.1.5 $a^2 - b^2 = (a + b)(a - b)$

**CD
1.1.5**

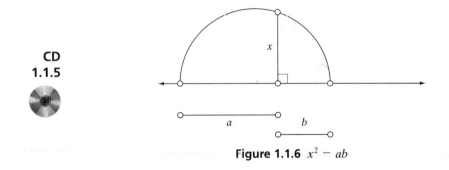

Figure 1.1.6 $x^2 = ab$

In addition to developing new mathematics, the Pythagoreans introduced sequences of propositions in which each successive assertion is justified by previous propositions. This practice shaped the development of both Greek mathematics and all subsequent mathematical research.

**URL
1.1.5**

Hippocrates of Chios (470–410 B.C.)

According to Proclus, Hippocrates of Chios was the first to actually compile an *Elements* of geometry. While no copies of this manuscript survive, many historians believe that Hippocrates' *Elements* was incorporated into the first two books of Euclid's *Elements.* Hippocrates taught in Athens and, among other things, worked on the classical problems of squaring the circle and duplicating the cube.

The term *squaring the circle* (also known as the *quadrature of the circle*) refers to the problem of finding a square with an area equal to that of a given circle. While we know today that this is impossible using only straight edge and compass, the ancient Greeks expended great effort attempting to do so using any means. To their credit, the ancient Greeks did not produce erroneous "proofs" of the quadrature of the circle. Lack of success was no impediment to their enthusiasm for this problem, however. Indeed, the problem became so popular that they invented a word meaning "to busy oneself with the quadrature."

While Hippocrates did not succeed in squaring the circle, he did develop methods for determining the areas of certain lunes, crescent-shaped areas created using arcs of circles. Of greater significance is Hippocrates' discovery that the ratio of the areas of two circles is the same as the ratio of the squares of their radii.

**URL
1.1.6**

Plato (427–347 B.C.)

Among the many remarkable philosophers and mathematicians of ancient times, relatively few shaped the development of western thought as powerfully as Plato. Born the youngest son of a wealthy Athenian family, Plato founded and directed the Academy, a school dedicated to research and instruction in philosophy and the sciences. At Plato's Academy students began by studying the exact sciences for ten years in order to prepare the mind to study relationships which only become clear when studied abstractly: Arithmetic, plane and solid geometry, astronomy, and harmonics. Following this introduction, five years would then be given to the study of "dialectic." Taken as the art of question and answer, Plato viewed dialectic as an essential skill in the investigation and demonstration of mathematical truth. By training young mathematicians how to prove propositions and test hypotheses, Plato helped to create a mathematical culture in which systematic progress was not only possible, but nearly guaranteed. Even Aristotle moved his school to Athens so that he and Plato might cooperate.

URL 1.1.7

While Plato's name is associated with the Platonic solids (cube, tetrahedron, octahedron, icosahedron, and dodecahedron), no important mathematical discoveries are attributed to Plato himself. All of the great mathematical discoveries of his time were made by his pupils. Closed in 529 A.D. by Emperor Justinian, Plato's Academy flourished for 900 years.

URL 1.1.8

Eudoxus (408–355 B.C.)

Eudoxus was a contemporary and a student of Plato. Most significant among his discoveries is a method for dealing with irrational numbers, a problem that had confounded mathematicians since the time of the Pythagoreans. Archimedes credits Eudoxus with proving two familiar propositions from solid geometry: The volume of a pyramid is one-third the volume of the prism having the same base and equal height; and the volume of a cone is one-third the volume of the cylinder having the same base and height. Eudoxus may also have created the first formal axiomatic system, based on Aristotle's theory of statements involving axioms, postulates, and definitions.

Summary

URL 1.1.9

Thales, Pythagoras, Hippocrates, Plato, and Eudoxus were part of a remarkable chain of Greek mathematicians. Figure 1.1.7 shows life times of these and other individuals who made meaningful contribution to the birth of mathematics in the western world. While this figure culminates in Archimedes, even he was but one link in a chain that spanned centuries. The individual links and the chain itself are the very foundation of our heritage as mathematicians.

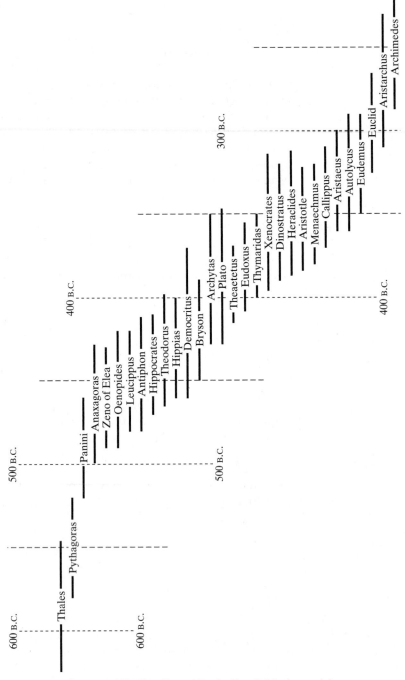

Figure 1.1.7 Timeline of Early Greek Mathematicians

URLs		Note: Begin each URL with the prefix http://
1.1.1	h	www-history.mcs.st-and.ac.uk/HistTopics/ Babylonian_and_Egyptian.html
1.1.2	h	www-history.mcs.st-and.ac.uk/ Indexes/Greek_index.html
1.1.3	h	www-history.mcs.st-and.ac.uk/ Mathematicians/Thales.html
1.1.4	h	www-history.mcs.st-and.ac.uk/ Mathematicians/Pythagoras.html
1.1.5	h	www-history.mcs.st-and.ac.uk/ Mathematicians/Hippocrates.html
1.1.6	h	www-history.mcs.st-and.ac.uk/ Mathematicians/Plato.html
1.1.7	s	www.ScienceU.com/geometry/facts/solids/handson.html
1.1.8	h	www-history.mcs.st-and.ac.uk/ Mathematicians/Eudoxus.html
1.1.9	h	www-history.mcs.st-and.ac.uk/ Chronology/ChronologyA.html

Table 1.1.1 Section 1.1 URLs (c = concept, h = history, s = software, d = data)

Exercises

1. Which culture had a more accurate method for computing the area of a circle, the Babylonians or the Egyptians? Why?
2. Solve this problem found on a Babylonian tablet: "An area A consisting of the sum of two squares is 1000. The side of one square is 10 less than 2/3 the side of the other square. What are the sides of the square?" Solve this problem using algebra as you learned it. Then imagine solving the problem before the invention of Algebra.
3. An inscription on an Egyptian papyrus states: "If you are told: A truncated pyramid of 6 for the vertical height, by 4 on the base, by 2 on the top. You are to square this 4, result 16. You are to double 4, result 8. You are to square 2, result 4. You are to add the 16, the 8, and the 4, result 28. You are to take one-third of 6, result 2. You are to take 28 twice, result 56. See, it is 56. You will find it right." What formula is being described? Write the formula using variables for the dimensions of the truncated pyramid.
4. To the naked eye, it appears that the area of a circle may be halfway between those of an inscribed square and a circumscribed square. If this observation is taken as true and accurate, what value must π take?
5. Pythagorean triples are positive integers a, b, and c that satisfy the Pythagorean equation $a^2 + b^2 = c^2$.

a) Find three sets of Pythagorean triples other than 3, 4, 5.
b) Prove that 3, 4, 5 is the only Pythagorean triple with consecutive integers as sides.
c) Prove that no isosceles right triangles exist for which the sides are Pythagorean triples.
d) Prove that in every Pythagorean triple, at least one leg is a multiple of three.

6. Explain Figures 1.1.3–1.1.6.
7. Show that the number $\sqrt{2}$ is irrational.
8. Using the Pythagorean method, solve the following equations for x:

   a) $x^2 = 12$

   b) $x^2 = 6$

9. Based on the following geometric model, prove that
   $(a + b)^2 > (a - b)^2.$

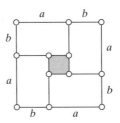

10. Use the following geometric model to find a formula for the sum of the first n odd counting numbers. Justify your answer.

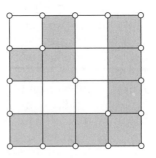

11. Find a geometric model for the sum of the first n counting numbers. Justify your answer.

1.2 Euclid and the *Elements*

In this section,
you will . . .
- **Investigate the content and organization of Euclid's *Elements*.**
- **Explore axiomatic systems.**
- **Investigate the structure of finite geometries.**

URLs
1.2.1
&
1.2.2
The great city of Alexandria, Egypt was founded in 332 B.C. by Alexander the Great. Located at the intersection of important trade routes, the city also became a crossroads of learning. Ptolemy I began building the University of Alexandria, the world's first great university, in 306 B.C. and opened the doors of the institution about 300 B.C. A Greek mathematician named Euclid was chosen to head the department of mathematics.

URL
1.2.3
Little is known about the personal life of Euclid (330–270 B.C.). According to Proclus,

> All those who have written histories bring to this point their account of the development of this science. Not long after these men came Euclid, who brought together the *Elements,* systematizing many of the theorems of Eudoxus, perfecting many of those of Theatetus, and putting in irrefutable demonstrable form propositions that had been rather loosely established by his predecessors. He lived in the time of Ptolemy the First, for Archimedes, who lived after the time of the first Ptolemy, mentions Euclid. It is also reported that Ptolemy once asked Euclid if there was not a shorter road to geometry than through the *Elements,* and Euclid replied that there was no royal road to geometry. He was therefore later than Plato's group but earlier than Eratosthenes and Archimedes, for these two men were contemporaries, as Eratosthenes somewhere says. Euclid belonged to the persuasion of Plato and was at home in this philosophy; and this is why he thought the goal of the *Elements* as a whole to be the construction of the so-called Platonic figures. (Morrow, 1970)

A prolific writer, Euclid wrote many books, of which the following survive: *Elements* (a systematic, axiomatic approach to geometry); *Data* (deducible properties of figures, given other properties); *On Divisions* (how to divide a figure into two parts with areas of a given ratio); *Optics* (the first Greek work on perspective); and *Phaenomena* (an elementary introduction to mathematical astronomy). Euclid's lost books include: *Surface Loci* (two books); *Porisms* (a three book work with, according to Pappus, 171 theorems and 38 lemmas); *Conics* (four books); *Book of Fallacies;* and *Elements of Music.*

Of his many works, Euclid's *Elements* may be, other than the Bible, the most published, translated, and studied book of Western civilization. More than one thousand editions of the *Elements* have been published since it was first printed in 1482. In writing the *Elements,* Euclid incorporated much of what was known of geometry at the time into a single mathematical framework. This synthesis provided both a context for previously discovered theorems and a means for identifying open unsolved problems. The main topics addressed in the *Elements* are listed in Table 1.2.1.

Book	Topic or Theme
1	Triangles, parallels, and area
2	Geometric algebra
3	Circles
4	Inscribed and circumscribed objects
5	Abstract proportions
6	Similarity
7	Fundamentals of number theory
8	Continued proportions in number theory
9	Number theory
10	Incommensurable (rational vs. irrational) numbers
11	Solid geometry
12	Measurement
13	Regular solids

Table 1.2.1 The Books of Euclid's *Elements*

These thirteen books include 465 propositions, or theorems. Traditional courses in high school geometry typically address much of the material found in Books 1, 3, 4, 6, 9, and 12. To millions of students whose schooling limited their geometrical training to these materials, Euclidean geometry *is* geometry, i.e. it is the only geometry. While it is an express goal of this text to enlarge student understanding of geometry to encompass more than Euclid imagined, no better entrance to the world of geometry exists than through Euclid's time-honored *Elements*.

In creating the *Elements*, Euclid did far more than produce a sort of encyclopedia of mathematics. He created the first system of thought based on formal definitions, axioms, propositions, and rules of inference, or logic. In doing so, Euclid established a standard for demonstrative mathematics that survives to this day. That standard is proof.

URL 1.2.4 The geometry of Euclid is based on defined terms and assumptions, called axioms or postulates. Using definitions and axioms as building blocks, Euclid demonstrated the truth of hundreds of other statements, called propositions. Table 1.2.2 illustrates his approach using selected examples of these elements of geometry from David Joyce's on-line version of *Euclid's Elements*.

Element	**Book**	**Examples** (Euclid's numbering)
Definition	I	1. A *point* is that which has no part.
		2. A *line* is breadthless length.
		3. The ends of a line are points.
		4. A *straight line* is a line which lies evenly with the points on itself.
		11. An *obtuse angle* is an angle greater than a right angle.
		12. An *acute angle* is an angle less than a right angle.
		21. Further, of trilateral figures, a *right-angled triangle* is that which has a right angle, an *obtuse-angled triangle* that which has an obtuse angle, and an *acute-angled triangle* that which has its three angles acute.
		22. Of quadrilateral figures, a *square* is that which is both equilateral and right-angled; an *oblong* that which is right-angled but not equilateral; a *rhombus* that which is equilateral but not right-angled; and a *rhomboid* that which has its opposite sides and angles equal to one another but is neither equilateral nor right-angled. And let quadrilaterals other than these be called *trapezia.*
		23. *Parallel* straight lines are straight lines which, being in the same plane and being produced indefinitely in both directions, do not meet one another in either direction.
Axioms	I	1. To draw a straight line from any point to any point.
		2. To produce a finite straight line continuously in a straight line.
		3. To describe a circle with any center and radius.
		4. That all right angles equal one another.

Table 1.2.2 Sample Definitions, Axioms, and Propositions

Element	Book	Examples (Euclid's numbering)
		5. That, if a straight line falling on two straight lines makes the interior angles on the same side less than two right angles, the two straight lines, if produced indefinitely, meet on that side on which are the angles less than the two right angles.
Common Notions	I	1. Things which equal the same thing also equal one another.
		2. If equals are added to equals, then the wholes are equal.
		3. If equals are subtracted from equals, then the remainders are equal.
		4. Things which coincide with one another equal one another.
		5. The whole is greater than the part.
Propositions	I	6. If in a triangle two angles equal one another, then the sides opposite the equal angles also equal one another.
		16. In any triangle, if one of the sides is produced, then the exterior angle is greater than either of the interior and opposite angles.
		17. In any triangle the sum of any two angles is less than two right angles.
		18. In any triangle the angle opposite the greater side is greater.
		19. In any triangle the side opposite the greater angle is greater.
		20. In any triangle the sum of any two sides is greater than the remaining one.

Table 1.2.2 Sample Definitions, Axioms, and Propositions *(continued)*

Axiomatic Systems

In general, axiomatic systems include the following elements or features: Undefined terms; defined terms; axioms; propositions; and rules of inference, or logic. Definitions are used to create a technical vocabulary for describing the features of objects, numbers, concepts and their relationships. For instance, Euclid defined an *acute angle* as an angle less than a right angle. In defining the term *right angle*, he stated: When a straight line standing on a straight line makes the adjacent angles equal to one another, each of the equal angles is right, and the straight line standing on the other is called a perpendicular to that on which it stands. This definition, in turn, makes use of the terms *adjacent* and *perpendicular.* While these terms may also be defined, repeated application of this process eventually forces one to use fundamental concepts such as *point* or *line.* It is at this point that both mathematics and language begin to falter. For instance, in attempting definitions for *point* and *line,* Euclid had to rely on vague notions and qualities that are not themselves defined: A *point* is that which has no part; and a *line* is breadthless length. Modern mathematicians recognize that this circumstance is an inevitable feature of mathematical logic and accommodate the situation by identifying the terms *point* and *line* as *undefined.* These terms are still useable in definitions, but no attempt is made to define them, thereby ending the chain of definitions at the point where further definition no longer adds additional information.

Axioms are unproven statements that are assumed to be true. For Euclid, his contemporaries, and for countless students of geometry through the ages, the following axioms, or postulates, were self-evident other than the fifth, which was in controversy for centuries.

1. To draw a straight line from any point to any point.
2. To produce a finite straight line continuously in a straight line.
3. To describe a circle with any center and radius.
4. That all right angles equal one another.
5. That, if a straight line falling on two straight lines makes the interior angles on the same side less than two right angles, the two straight lines, if produced indefinitely, meet on that side on which are the angles less than the two right angles (see Figure 1.2.1).

CD
1.2.1

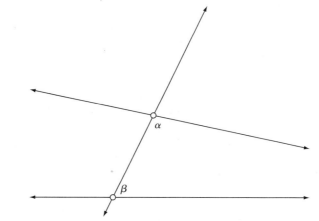

Figure 1.2.1 Euclid's Fifth Axiom

In most U.S. high school geometry texts, axioms 1–4 are restated using contemporary language and the fifth axiom is replaced by a different statement by John Playfair (see Figure 1.2.2) that is logically equivalent to Euclid's fifth axiom:

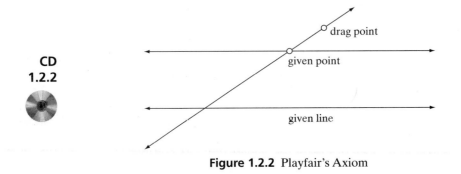

**CD
1.2.2**

Figure 1.2.2 Playfair's Axiom

1. A line may be drawn between any two points.
2. A segment of any length may be constructed in any line.
3. A circle with any radius and center may be drawn.
4. All right angles are equal.
5. Through a point not on a given line, exactly one line may be drawn parallel to the given line.

Propositions are statements that are proven to be true using accepted rules of inference. In the *Elements,* propositions are stated using both declarative and If-Then sentence formats. For example, the proposition "To draw a straight line at right angles to a given straight line from a given point on it" declares that a perpendicular may be drawn to a line from a point on the line. In the proposition "If two straight lines cut one another, then they make the vertical angles equal to one another," the If portion of the statement and the Then portion of the statement specify a relationship between a given set of conditions and a particular conclusion or consequence. One of the most significant features of the *Elements* is the manner in which sequences of propositions lead to progressively broader and deeper insights. This approach made it possible for geometers to catalog the scope of their knowledge and to identify significant unsolved, or open, problems.

**URL
1.2.5**
The rules of inference, or mathematical logic, used by Euclid were formalized by the Greek philosopher Aristotle (384–322 B.C.). Aristotle was a student of Plato and taught at the Academy for many years. More than any other philosopher, Aristotle shaped the orientation and subsequent evolution of Western intellectual history. The author of the "scientific method," his ideas permeate Western thinking. Known principally as a philosopher rather than as a mathematician, Aristotle was interested in the logical structure of mathematical and scientific arguments. Consequently, questions such as "What conditions must a statement satisfy to qualify as an axiom?" and "By what rules should propositions be deduced from axioms?" served as the focus of his mathematical writings.

Today, mathematicians describe axiomatic systems using concepts that have been developed over centuries. The concepts of consistency, independence, and completeness are fundamental to such studies. These concepts are normally introduced using models of finite geometries.

**Definition
1.2.1**

An *axiomatic system* is said to be consistent if neither the axioms nor the propositions of the system contradict one another.

**Definition
1.2.2**

A *axiom* is said to be independent of the other axioms if it cannot be derived from the other axioms.

**Definition
1.2.3**

An axiomatic system is *complete* if it is possible for every properly posed statement to be proved or disproved. Alternatively, it is not possible to add a new independent axiom to the system.

Modeling Axiomatic Systems Using Finite Geometries

A finite geometry consists of a finite set of points and an associated set of terms and axioms. Figure 1.2.3 shows a model for Four Point geometry. Because this is a finite geometry, normal Euclidean lines do not exist. Nevertheless, it is possible to define sets of points and call the sets lines. In this case, lines consist of pairs of points, represented in Figure 1.2.3 by lines in the figure and by columns of letters in the model.

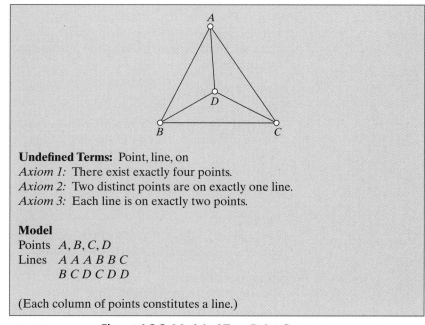

Undefined Terms: Point, line, on
Axiom 1: There exist exactly four points.
Axiom 2: Two distinct points are on exactly one line.
Axiom 3: Each line is on exactly two points.

Model
Points A, B, C, D
Lines $A\ A\ A\ B\ B\ C$
  $B\ C\ D\ C\ D\ D$

(Each column of points constitutes a line.)

Figure 1.2.3 Model of Four Point Geometry

This model provides a convenient context in which to demonstrate the concept of consistency. To do so, each axiom is verified, one at a time. There are exactly four points. Given any two points, there is only one line containing the two points. Given any line, it contains exactly two points.

Demonstrating independence is somewhat more difficult, both conceptually and procedurally. Two insights are particularly useful in considering the question of independence. First, if a given set of consistent axioms is independent, and if one axiom is negated while the others remain unchanged, the axiomatic system that results must also be independent and should be demonstrable in some consistent model, though not necessarily the same model used to illustrate the original system. Second, if a given set of axioms is dependent, and if one axiom is negated while the others remain unchanged, the axiomatic system that results must also be dependent.

In the case of Four Point geometry, demonstrating the independence of the three axioms involves negating one axiom at a time (see Example 1.2.1).

Example 1.2.1   Show that the axioms for Four Point geometry are independent.

Negation of Axiom 1:
 Axiom 1: There exist exactly two points.
 Axiom 2: Two distinct points are on exactly one line.
 Axiom 3: Each line is on exactly two points.
 Model
 Points A, B
 Lines A
   B
 Verification
 The two points are on exactly one line and the only line is on those two points.

Negation of Axiom 2:
 Axiom 1: There exist exactly four points.
 Axiom 2: Two distinct points are not on exactly one line.
 Axiom 3: Each line is on exactly two points.
 Model
 Points A, B, C, D
 Lines $A\ C$
   $B\ D$
 Verification
 There are exactly four points, two points (B and C) are not on any line, and each line is on exactly two points.

Negation of Axiom 3:
 Axiom 1: There exist exactly four points.
 Axiom 2: Two distinct points are on exactly one line.
 Axiom 3: Each line is not on exactly two points.
 Model
 Points A, B, C, D
 Lines $A\ A\ B\ C$
   $B\ D\ D\ D$
   C
 Verification
 There are exactly four points, each pair of points is on exactly one line, and each line is on exactly two points.

URLs
1.2.6
&
1.2.7
Depending on the axiomatic system under study, finding models for each negation may be relatively straightforward, as in Example 1.2.1, or very challenging. Investigations of the latter sort are normally undertaken by specialists in mathematical logic. Among modern logicians, the most famous is Kurt Gödel (1906–1978). Gödel is best known for proving in 1931 that, in any axiomatic mathematical system, rich enough to encompass the axioms of arithmetic, there are propositions that cannot be proved or disproved within the axioms of the system. This is known as Gödel's Incompleteness Theorem. An interesting question in light of this finding is whether a computer could ever be programmed to answer all mathematical questions. What do you think? Relative to Definition 1.2.3, Gödel showed that most "interesting" axiomatic systems, such as arithmetic and geometry, are incomplete.

Summary

In creating the *Elements*, Euclid achieved a fundamental breakthrough in the development of mathematics that has shaped the development of Western intellectual history more profoundly than any other scholarly work. In the *Elements*, Euclid did more than synthesize what was then known about geometry. He developed a system of thought that serves as the infrastructure for development of all subsequent mathematics, the mathematical logic of axiomatic systems. For over two thousand years, educated people have viewed the *Elements* as the natural pedagogical instrument for introducing students to deductive reasoning and axiomatic systems. Axiomatic reasoning is used extensively in all branches of mathematics, computer science, and the sciences. More than ever, the *Elements* offers all who would take up a study of mathematical logic a time-honored introduction to geometry.

Definitions	
1.2.1	An axiomatic system is said to be *consistent* if neither the axioms nor the propositions of the system contradict one another.
1.2.2	An axiom is said to be *independent* of the other axioms if it cannot be derived from the other axioms.
1.2.3	An axiomatic system is *complete* if it is possible for every properly posed statement to be proved valid or invalid. Alternatively, it is not possible to add a new independent axiom to the system.

Table 1.2.3 Summary, Section 1.2

URLs		Note: Begin each URL with the prefix http://
1.2.1	h	aleph0.clarku.edu/~djoyce/mathhist/alexandria.html
1.2.2	h	www.perseus.tufts.edu/GreekScience/Students/Ellen/Museum.html
1.2.3	h	www-history.mcs.st-and.ac.uk/history/Mathematicians/Euclid.html
1.2.4	c	aleph0.clarku.edu/~djoyce/java/elements/toc.html
1.2.5	h	www-history.mcs.st-and.ac.uk/history/Mathematicians/Aristotle.html
1.2.6	h	www-history.mcs.st-and.ac.uk/history/Mathematicians/Godel.html
1.2.7	h	users.ox.ac.uk/~jrlucas/mmg.html

Table 1.2.4 Section 1.2 URLs (c = concept, h = history, s = software, d = data)

Exercises

1. Write each of the following statements from the *Elements* in your own words.

   a) A circle is a plane figure contained by one line such that all the straight lines falling upon it from one point among those lying within the figure equal one another. (Definition 15)

   b) A diameter of the circle is any straight line drawn through the center and terminated in both directions by the circumference of the circle, and such a straight line also bisects the circle. (Definition 17)

   c) Of trilateral figures, an equilateral triangle is that which has its three sides equal, an isosceles triangle that which has two of its sides alone equal, and a scalene triangle that which has its three sides unequal. (Definition 20)

   d) Of trilateral figures, a right-angled triangle is that which has a right angle, an obtuse-angled triangle that which has an obtuse angle, and an acute-angled triangle that which has its three angles acute. (Definition 21)

   e) Of quadrilateral figures, a square is that which is both equilateral and right-angled; an oblong that which is right-angled but not equilateral; a rhombus that which is equilateral but not right-angled; and a rhomboid that which has its opposite sides and angles equal to one another but is neither equilateral nor right-angled. And let quadrilaterals other than these be called trapezia. (Definition 22)

   f) Parallel straight lines are straight lines which, being in the same plane and being produced indefinitely in both directions, do not meet one another in either direction. (Definition 23)

   g) If two triangles have two sides equal to two sides respectively, and have the angles contained by the equal straight lines equal, then they also have the base equal to the base, the triangle equals the triangle, and the remaining angles equal the remaining angles respectively, namely those opposite the equal sides. (Proposition 4)

h) In isosceles triangles the angles at the base equal one another, and, if the equal straight lines are produced further, then the angles under the base equal one another. (Proposition 5)

i) In any triangle the sum of any two angles is less than two right angles. (Proposition 17)

j) If two triangles have two angles equal to two angles respectively, and one side equal to one side, namely, either the side adjoining the equal angles, or that opposite one of the equal angles, then the remaining sides equal the remaining sides and the remaining angle equals the remaining angle. (Proposition 26)

2. Make a sketch to illustrate each definition or proposition in exercise #1.

3. Using the historical resources available at URL 1.2.1 and 1.2.2, investigate the political and intellectual context in which Euclid worked at the University of Alexandria, Egypt. Summarize your conclusions.

4. Using a Java-enabled WWW browser visit David Joyce's on-line version of the *Elements* at URL 1.2.4. Review Euclid's proofs of the following propositions, then rewrite the proofs using contemporary English.

a) Book I, Proposition 17
b) Book I, Proposition 26
c) Book I, Proposition 41

5. Three Point geometry is defined as follows:

Undefined terms: Point, line, on

Axiom 1: There exist exactly three points.

Axiom 2: Every pair of points is on exactly one line.

Axiom 3: Every pair of lines is on at least one point.

Axiom 4: No three points are collinear.

a) Is this system consistent?

b) Is the system independent?

[Note: This is an example of a complete axiomatic system.]

1.3 Neutral Geometry

In this section you will . . .

- **Investigate neutral geometry through a study of the first 28 Propositions of Euclid's *Elements*.**

The foundation of Euclidean geometry is expressed in its axioms:

1. To draw a straight line from any point to any point.
2. To produce a finite straight line continuously in a straight line.
3. To describe a circle with any center and radius.
4. That all right angles equal one another.
5. That, if a straight line falling on two straight lines makes the interior angles on the same side less than two right angles, the two straight lines, if produced indefinitely, meet on that side on which are the angles less than the two right angles.

While all of Euclidean geometry is based on these axioms, and a limited number of unstated axions discovered later, some propositions may be proven using only axioms 1–4. For instance, Propositions 1–28 of Book I of the *Elements* make no reference (directly or indirectly) to the fifth axiom. To many geometers, this approach suggested that the fifth axiom might be dependent on the first four axioms and therefore provable as a proposition. The search for such a proof motivated, challenged, and frustrated students of geometry for over two thousand years. Ultimately, that search led to the discovery of non-Euclidean geometries (see Chapter Three), each of which involves modifications to axioms 1–4 and a negation of Euclid's fifth axiom.

Neutral geometry uses neither Euclid's fifth axiom nor its possible negations. With modifications, many of the propositions of neutral geometry are true in both Euclidean and non-Euclidean geometry. Consequently, neutral geometry provides a convenient frame of reference with which to compare and contrast the features of Euclidean and non-Euclidean geometries. It is also a natural starting point for the study of Euclidean geometry.

Definition 1.3.1

A plane geometry is *neutral* if it does not include a parallel postulate or its logical consequences.

The First 28 Propositions of Book I of Euclid's *Elements*
The following theorems illustrate the content and format of Euclid's first 28 propositions.

Theorem 1.3.1

If in a triangle two angles equal one another, then the sides opposite the equal angles also equal one another (Proposition 6; see Figure 1.3.1).

URLs 1.3.1 & 1.3.2

Proposition 6 presents the first "proof by contradiction" in the *Elements*. Figure 1.3.1 shows Euclid's proof as illustrated on David Joyce's WWW site Euclid's *Elements*. The illustration in Figure 1.3.1 is interactive, enabling the user to reposition points A, B, C, and D. Euclid's argument is presented in paragraph format. Notations in the right hand column refer

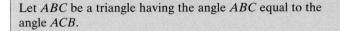

If in a triangle two angles equal one another, then the sides opposite the equal angles also equal one another.

Let *ABC* be a triangle having the angle *ABC* equal to the angle *ACB*.

I say that the side *AB* also equals the side *AC*.

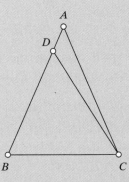

If *AB* does not equal *AC*, then one of them is greater. C.N.

Let *AB* be greater. Cut off *DB* from *AB* the greater equal   I.3
to *AC* the less, and join *DC*.   Post.1

Since *DB* equals *AC*, and *BC* is common, therefore the two
sides *DB* and *BC* equal the two sides *AC* and *CB* respectively, and the angle *DBC* equals the angle *ACB*. Therefore   I.4
the base *DC* equals the base *AB*, and the triangle *DBC*   C.N.5
equals the triangle *ACB*, the less equals the greater, which
is absurd. Therefore *AB* is not unequal to *AC*, it therefore
equals it.

Figure 1.3.1 Proposition 6 *Adapted from Euclid's Elements
by David Joyce (see URL 1.3.1).*

to Common Notions (C.N.), Propositions (I.), and Postulates (Post.). Clicking on any of these notations takes the user to the justifying Common Notion, Proposition, Postulate, or Definition. Notice that no reference is made to Euclid's fifth axiom either in the proof of Proposition 6 or in any notations. The proof of Proposition 6 may be rephrased using contemporary English as follows:

Let *ABC* be a triangle in which $\angle ABC = \angle ACB$. Assume that $AB \neq AC$. Let side *AB* be longer than side *AC*. Locate point *D* on side *AB* such that $DB = AC$. Then $\triangle DBC \cong \triangle ACB$ by side-angle-side (Proposition 4). Consequently, $DB = AB = AC$, which is a contradiction. Therefore if in a triangle two angles equal one another, then the sides opposite the equal angles are also equal.

Theorem 1.3.2   To bisect a given finite straight line (Proposition 10).

URL   Proposition 10 demonstrates how to construct the midpoint of a seg-
1.3.3   ment (see Figure 1.3.2). The strategy is based on proving that
$\triangle ACD \cong \triangle BCD$. Consequently $AD = BD$.

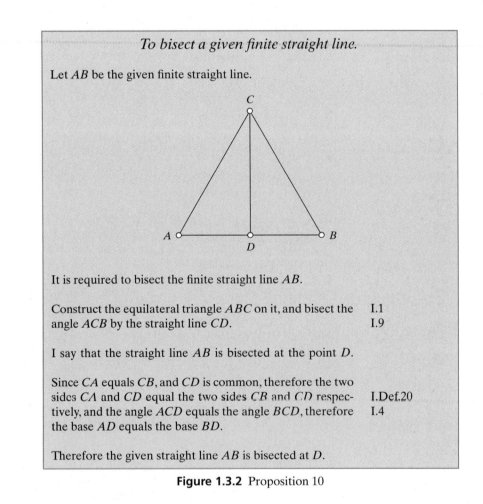

> ## To bisect a given finite straight line.
>
> Let AB be the given finite straight line.
>
> It is required to bisect the finite straight line AB.
>
> Construct the equilateral triangle ABC on it, and bisect the   I.1
> angle ACB by the straight line CD.   I.9
>
> I say that the straight line AB is bisected at the point D.
>
> Since CA equals CB, and CD is common, therefore the two
> sides CA and CD equal the two sides CB and CD respec-   I.Def.20
> tively, and the angle ACD equals the angle BCD, therefore   I.4
> the base AD equals the base BD.
>
> Therefore the given straight line AB is bisected at D.

Figure 1.3.2 Proposition 10

Theorem 1.3.3   In any triangle, if one of the sides is produced, then the exterior angle is
greater than either of the interior and opposite angles (Proposition 16;
see Figure 1.3.3).

Euclid's first 28 propositions appear in Tables 1.3.1–1.3.3 on pages 26–28.

In any triangle, if one of the sides is produced, then the exterior angle is greater than either of the interior and opposite angles.

Let *ABC* be a triangle, and let one side of it *BC* be produced to *D*.

I say that the exterior angle *ACD* is greater than either of the interior and opposite angles *CBA* and *BAC*.

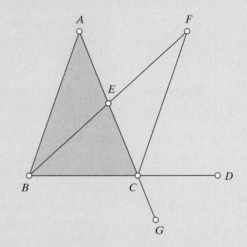

Bisect *AC* at *E*. Join *BE* and produce it in a straight line to *F*.   I.10, Post.1
                                                                                 Post.2

Make *EF* equal to *BE*, join *FC*, and draw *AC* through to *G*.   I.3, Post.1
                                                                           Post.2

Since *AE* equals *EC*, and *BE* equals *EF*, therefore the two sides *AE* and *EB* equal the two sides *CE* and *EF* respectively, and the angle *AEB* equals the angle *FEC* for they are vertical angles. Therefore the base *AB* equals the base   I.15
FC, and triangle *ABE* equals the triangle *CFE*, and the   I.4
remaining angles equal the remaining angles respectively, namely those opposite the equal sides. Therefore the angle *BAE* equals the angle *ECF*.

But the angle *ECD* is greater than the angle *ECF*, therefore the angle *ACD* is greater than the angle *BAE*.   C.N.5

Similarly, if *BC* is bisected, then the angle *BCG*, that is, the angle *ACD*, can also be proved to be greater than the angle   I.15
ABC.

Therefore *in any triangle, if one of the sides is produced, then the exterior angle is greater than either of the interior and opposite angles.*

Figure 1.3.3 Proposition 16

Proposition 1	To construct an equilateral triangle on a given finite straight line.
Proposition 2	To place a straight line equal to a given straight line with one end at a given point.
Proposition 3	To cut off from the greater of two given unequal straight lines a straight line equal to the less.
Proposition 4	If two triangles have two sides equal to two sides respectively, and have the angles contained by the equal straight lines equal, then they also have the base equal to the base, the triangle equals the triangle, and the remaining angles equal the remaining angles respectively, namely those opposite the equal sides.
Proposition 5	In isosceles triangles the angles at the base equal one another, and, if the equal straight lines are produced further, then the angles under the base equal one another.
Proposition 6	If in a triangle two angles equal one another, then the sides opposite the equal angles also equal one another.
Proposition 7	Given two straight lines constructed from the ends of a straight line and meeting in a point, there cannot be constructed from the ends of the same straight line, and on the same side of it, two other straight lines meeting in another point and equal to the former two respectively, namely each equal to that from the same end.
Proposition 8	If two triangles have the two sides equal to two sides respectively, and also have the base equal to the base, then they also have the angles equal which are contained by the equal straight lines.
Proposition 9	To bisect a given rectilinear angle.
Proposition 10	To bisect a given finite straight line.
Proposition 11	To draw a straight line at right angles to a given straight line from a given point on it.
Proposition 12	To draw a straight line perpendicular to a given infinite straight line from a given point not on it.
Proposition 13	If a straight line stands on a straight line, then it makes either two right angles or angles whose sum equals two right angles.
Proposition 14	If with any straight line, and at a point on it, two straight lines not lying on the same side make the sum of the adjacent angles equal to two right angles, then the two straight lines are in a straight line with one another.

Table 1.3.1 Propositions 1–14

Proposition 15	If two straight lines cut one another, then they make the vertical angles equal to one another.
Proposition 16	In any triangle, if one of the sides is produced, then the exterior angle is greater than either of the interior and opposite angles.
Proposition 17	In any triangle the sum of any two angles is less than two right angles.
Proposition 18	In any triangle the angle opposite the greater side is greater.
Proposition 19	In any triangle the side opposite the greater angle is greater.
Proposition 20	In any triangle the sum of any two sides is greater than the remaining one.
Proposition 21	If from the ends of one of the sides of a triangle two straight lines are constructed meeting within the triangle, then the sum of the straight lines so constructed is less than the sum of the remaining two sides of the triangle, but the constructed straight lines contain a greater angle than the angle contained by the remaining two sides.
Proposition 22	To construct a triangle out of three straight lines which equal three given straight lines: thus it is necessary that the sum of any two of the straight lines should be greater than the remaining one.
Proposition 23	To construct a rectilinear angle equal to a given rectilinear angle on a given straight line and at a point on it.
Proposition 24	If two triangles have two sides equal to two sides respectively, but have one of the angles contained by the equal straight lines greater than the other, then they also have the base greater than the base.
Proposition 25	If two triangles have two sides equal to two sides respectively, but have the base greater than the base, then they also have one of the angles contained by the equal straight lines greater than the other.
Proposition 26	If two triangles have two angles equal to two angles respectively, and one side equal to one side, namely, either the side adjoining the equal angles, or that opposite one of the equal angles, then the remaining sides equal the remaining sides and the remaining angle equals the remaining angle.

Table 1.3.2 Propositions 15–26

Proposition 27	If a straight line falling on two straight lines makes the alternate angles equal to one another, then the straight lines are parallel to one another.
Proposition 28	If a straight line falling on two straight lines makes the exterior angle equal to the interior and opposite angle on the same side, or the sum of the interior angles on the same side equal to two right angles, then the straight lines are parallel to one another.

Table 1.3.3 Propositions 27–28

Summary

In creating the *Elements*, Euclid began with 28 neutral propositions, that is, propositions that make no reference to the fifth axiom. While these propositions do not constitute all of neutral geometry, they do illustrate many elements of its features.

Table 1.3.4 summarizes the definitions and theorems discussed in this section.

Definitions	
1.3.1	A geometry is *neutral* if it does not include a parallel axiom or its logical consequences.
Theorems	
1.3.1	If in a triangle two angles equal one another, then the sides opposite the equal angles also equal one another (Proposition 6).
1.3.2	To bisect a given finite straight line (Proposition 10).
1.3.3	In any triangle, if one of the sides is produced, then the exterior angle is greater than either of the interior and opposite angles (Proposition 16).

Table 1.3.4 Summary, Section 1.3

URLs		Note: Begin each URL with the prefix http://
1.3.1	hc	aleph0.clarku.edu/~djoyce/java/elements/bookI/propI6.html
1.3.2	hc	aleph0.clarku.edu/~djoyce/java/elements/elements.html
1.3.3	hc	aleph0.clarku.edu/~djoyce/java/elements/bookI/propI10.html
1.3.4	hc	aleph0.clarku.edu/~djoyce/java/elements/bookI/propI16.html

Table 1.3.5 Section 1.3 URLs (c = concept, h = history, s = software, d = data)

Exercises

1. Rephrase each of the following propositions using contemporary English.
   a) Proposition 1
   b) Proposition 8
   c) Proposition 12
   d) Proposition 15
   e) Proposition 18
   f) Proposition 27
2. Rephrase Euclid's proofs of the following propositions using contemporary English.
   a) Proposition 1
   b) Proposition 8
   c) Proposition 12
   d) Proposition 15
   e) Proposition 18
   f) Proposition 27
3. Prove the following propositions of neutral geometry (i.e., without using Euclid's fifth axiom or its logical consequences).
   a) Every line segment has exactly one midpoint.
   b) Every angle has exactly one bisector.
   c) Supplements (or complements) of the same angle are congruent.
   d) If a line l intersects $\triangle ABC$ at a point P on segment BC, then l also intersects segment AC and/or AB (Pasch's Axiom).
   e) If Q is a point in the interior of $\triangle ABC$, then ray AQ intersects segment BC at a point D (Crossbar Theorem).
   f) Equilateral triangles are also equiangular.

1.4   Famous Problems In Geometry

In this section, you will . . .

• **Investigate three famous problems that were introduced in the time of the Greeks and which remained unsolved for over 2000 years.**

In every branch of mathematics, advancement is often associated with progress with respect to *open* (well-defined but unsolved) problems. In the history of geometry, three problems remained open for two thousand years, attracting the attention of the best geometers of each age. Each of these problems calls for a construction using only straight edge and compass. While the greatest of geometers were unable to achieve the indicated results using only straightedge and compass (something we know today to be impossible), the problems repeatedly drew outstanding geometers into dialogues that were fruitful in other ways.

Trisecting an Angle

Using only straightedge and compass, trisect a given angle (see Figure 1.4.1).

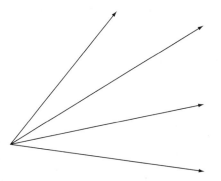

Figure 1.4.1 Trisecting an Angle

Certain angles are easy to trisect. For instance, trisecting a 90° angle is simple if you can construct a 30° angle. This may be achieved in two steps. First, construct an equilateral triangle. This produces angles of 60°. One of these 60° angles is then bisected to obtain a 30° angle. Can you think of other angles that may be trisected using similar strategies (see exercises)?

**CD
1.4.1**

The challenge in this problem is to trisect any given angle. Figures 1.4.2 and 1.4.3 illustrate a misconception demonstrated by many students considering the problem for the first time. Rays drawn from the vertex of the given angle through the trisection points of a segment intersecting the given angle do not trisect the given angle. If the angle to be trisected is small, and if the segment forms equal angles with the sides of the given angle, the given angle is divided into three angles that are approximately equal, but never exactly equal. If the angle to be trisected is large, the given angle is divided into three angles that are clearly different in size.

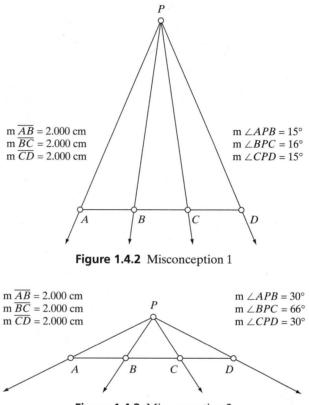

m $\overline{AB}$ = 2.000 cm

m $\overline{BC}$ = 2.000 cm

m $\overline{CD}$ = 2.000 cm

m $\angle APB$ = 15°

m $\angle BPC$ = 16°

m $\angle CPD$ = 15°

Figure 1.4.2 Misconception 1

m $\overline{AB}$ = 2.000 cm

m $\overline{BC}$ = 2.000 cm

m $\overline{CD}$ = 2.000 cm

m $\angle APB$ = 30°

m $\angle BPC$ = 66°

m $\angle CPD$ = 30°

Figure 1.4.3 Misconception 2

**URL
1.4.1**

    The first known "solution" to the problem of trisecting an angle is attributed to Hippocrates. The method of Hippocrates is illustrated in Figure 1.4.4. A perpendicular is drawn from point C on one side of given angle CAB to point D on the other side. Rectangle $CDAF$ is then constructed. Ray FC is then drawn and point E located on FC such that $HE = 2AC$. Under this condition, $\angle EAB = 1/3\angle CAB$.

**CD
1.4.2**

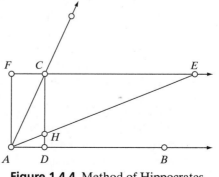

Figure 1.4.4 Method of Hippocrates

What about this approach violates the "straightedge and compass only" restriction of the original problem statement? Why does this approach trisect? (See exercises.)

Squaring the Circle

Using only ruler and compass, construct a square
equal in area to a given circle. (See Figure 1.4.5.)

Figure 1.4.5 Squaring the Circle

URL 1.4.2 This is a very, very old problem, appearing in one form or another in some of the world's oldest mathematical records. For instance, the Rhind papyrus, copied around 1650 B.C. by the Egyptian scribe Ahmes (1680 B.C.–1620 B.C.) from an even older document, gives a rule for constructing a square of area equal to that of a circle. The rule is to cut 1/9 off the circle's diameter and to construct a square on the remainder. While this approach is a "straightedge and compass only" construction, it does not actually produce the desired result but a very good approximation. This approach also provides a surprisingly good approximation for the value of π, 3.1605.

URL 1.4.3 The first mathematician known to have attempted squaring the circle is Anaxagoras (499 B.C.– 428 B.C.). Plutarch, in his work *On Exile* which was written in the first century A.D., says "There is no place that can take away the happiness of a man, nor yet his virtue or wisdom. Anaxagoras, indeed, wrote on the squaring of the circle while in prison."

URL 1.4.4 To the credit of the ancient Greeks, there is no record of fallacious "proofs" that the circle may be squared using only straightedge and compass. Once they realized that a proof of this sort was probably impossible, they sought other mathematical strategies for squaring the circle. For instance, Archimedes used the spiral that bears his name to create right triangles with the same area as given circles. In Figure 1.4.6, circle *AB* and the spiral have a common center. The radius of circle *AB* is drawn from *A* to the point at which the spiral completes 360° of its path, point *B*. Next, a tangent is drawn to the spiral at point *B* and a perpendicular to radius *AB* at point *A*. These lines intersect at point *C*. In Proposition 19 of *On Spirals*, Archimedes proved that the area of $\triangle ABC$ is equal to the area of circle *AB*. Finding a square with the same area is relatively simple (see exercises).

From a modern point of view, if a square with side *s* and a circle with radius *r* have identical areas, $s^2 = \pi r^2 \Rightarrow s = r\sqrt{\pi}$. Consequently, any procedure for squaring the circle must necessarily be capable of constructing the square root of π. Since π is a transcendental number, it is not the root of a rational polynomial equation. The implications of this fact are discussed later in this section.

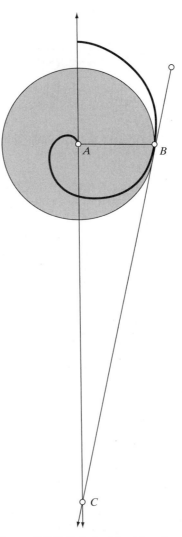

Figure 1.4.6 Spiral of Archimedes

Doubling the Cube

Using only ruler and compass, construct a
cube double in volume to a given cube.

Figure 1.4.7 demonstrates a simple method for doubling the area of any square. Using the Pythagorean Theorem, it is simple to show that the edge of the outside square is equal to the edge of the shaded square times the square root of two (see exercises). Figure 1.4.8 illustrates a common misconception: Doubling the side of the square does not double the area of the square. Instead, it leads to an area four times that of the original square.

URL
1.4.5   A similar misconception appears to be linked to the origin of the problem of doubling the cube (see Figure 1.4.9). In a tale told by Eratosthenes, around 430 B.C. a god demanded that the Delians construct

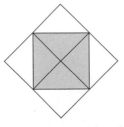

Figure 1.4.7 Doubling the
Square Solution

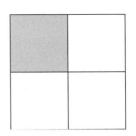

Figure 1.4.8 Doubling the
Square Misconception

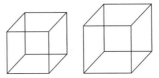

Figure 1.4.9 Doubling the Cube

an altar double in volume to that of the existing one … or suffer a terrible plague. When the craftsmen couldn't determine the new measurements, they went to Plato for advice. Plato is said to have replied that the real purpose of the god in setting them the task was to shame the Greeks for their neglect of mathematics and their contempt of geometry. We don't know their reaction, but legend states that shortly after a plague killed about a quarter of the population. The misconception is still alive and well, however. Many people believe that doubling the edge of the cube doubles its volume, when in fact the volume is multiplied by a factor of eight.

From a modern perspective, if the volume of one cube is given by s^3 and another by $2s^3$, the edges of the two cubes are related by the factor $\sqrt[3]{2}\,s$. Consequently, whether or not the cube may be doubled depends on whether the number $\sqrt[3]{2}$ may be constructed using straightedge and compass.

Constructible Numbers

**URL
1.4.6**

The question of whether a given construction is possible using only straightedge and compass was clearly answered and proven for the first time by Pierre Wantzel in 1837 after more than two thousand years as an open question. The following discussion presents a few of the principal concepts involved in resolving this ancient question.

**Definition
1.4.1**

A number a is *constructible* if one may construct a line segment with length $|a|$ in a finite number of steps from a segment of unit length using only a straightedge and compass.

**Theorem
1.4.1**

If a and b are constructible real numbers, then so are $a + b$, $a - b$, ab, a/b, and $\sqrt{a}$.

a) Since a and b are both constructible numbers, they may be represented as line segments. Placing these segments end-to-end along a line (see Figure 1.4.10) spans a distance equal to the sum of their lengths, $a + b$.

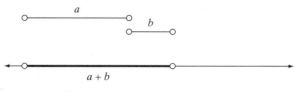

Figure 1.4.10 Constructing $a + b$

b) Since a and b are both constructible numbers, they may be represented as line segments. Overlapping these segments along a line as seen in Figure 1.4.11 directly demonstrates the difference in their lengths, $a - b$.

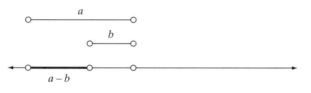

Figure 1.4.11 Constructing $a - b$

c) Since a and b are both constructible numbers, they may be represented as line segments. Using an arbitrary angle DAB as a scaffold (see Figure 1.4.12), segments of length a and b are positioned on opposite sides of the angle at the vertex, A. The other ends of segments a and b correspond to points B and D, respectively. In addition, a point C is located one unit from point A on ray AD. Segment CD is then drawn. Segment DE is drawn through point D parallel to segment CD. Using similar triangles, it is a simple matter to show that segment AE has length ab (see exercises).

CD
1.4.3

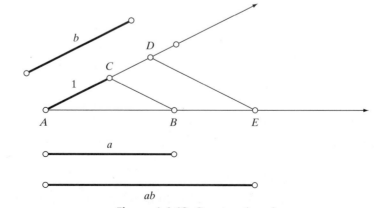

Figure 1.4.12 Constructing ab

d) Since *a* and *b* are both constructible numbers, they may be represented as line segments. Using an arbitrary angle *DAB* as a scaffold (see Figure 1.4.13), segments of length *a* and *b* are positioned on opposite sides of the angle at the vertex, *A*. The other ends of segments *a* and *b* correspond to points *E* and *D*, respectively. In addition, a point *C* is located one unit from point *A* on ray *AD*. Segment *DE* is then drawn. Segment *CB* is drawn through point *C* parallel to segment *DE*. Using similar triangles, it is a simple matter to show that segment *AB* has length a/b.

CD

1.4.4

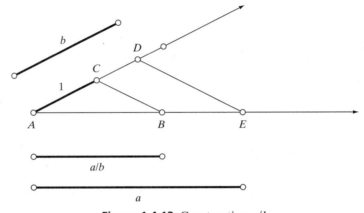

Figure 1.4.13 Constructing a/b

e) Since *a* and 1 are both constructible numbers, they may be represented as line segments. Segments of length *a* and 1 are positioned on opposite sides of point *B* (see Figure 1.4.14). The other ends of the segments correspond to points *A* and *C*, respectively. The midpoint of segment *AC* is constructed and used as the center of a semicircle. A perpendicular line is drawn from point *B* to point *D* on the semicircle. Using similar triangles, it is a simple matter to show that segment *DB* has length $\sqrt{a}$.

CD

1.4.5

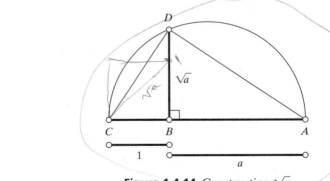

Figure 1.4.14 Constructing $\sqrt{a}$

The geometric tools of straightedge and compass are used to draw straight lines and circles. Straight lines have equations of the form

$ax + by + c = 0$. Circles have equations of the form $x^2 + y^2 + dx + ey + f = 0$. Geometric constructions focus on the intersections of lines with lines, lines with circles, and circles with circles. Algebraically, this corresponds to finding simultaneous solutions to systems of linear and/or quadratic equations.

The solutions to such systems are always expressed in terms of linear or quadratic equations, never cubic or higher order equations. For instance, solving the linear equation above for y yields $y = (-ax - c)/b$. When this is substituted into the equation of the circle, a quadratic equation in x is obtained: $x^2(a^2 + b^2) + x(2ac + b^2d - abe) + (c^2 - bec + b^2) = 0$. The roots of this equation may be obtained using the quadratic formula, which expresses solutions using square roots, not cubic or other roots. Every system of intersecting lines and circles produces the same sorts of solutions, rational numbers and/or square roots of rational numbers.

Consequently, all rational numbers are constructable, as well as their square roots, fourth roots (square roots of square roots), eight roots, and so on. While a full discussion on this topic is beyond the scope of this book, thorough treatments of the subject are readily available in many modern algebra texts.

**Theorem
1.4.2**   The field of constructible real numbers consists of all real numbers obtained by taking square roots of positive numbers a finite number of times and applying a finite number of field operations.

**Example
1.4.1**   Figure 1.4.15 shows that an angle α is constructible if and only if a segment of length $|\cos \alpha|$ is constructible. Similarly, an angle 3α is constructible if and only if a segment of length $|\cos 3\alpha|$ is constructible. Consequently, an angle 3α is trisectable if and only if a segment of length $|\cos \alpha|$ is constructible. Since $\cos 3\alpha = \cos(2\alpha + \alpha)$, using a familiar trigonometry identity, we may write

$$\cos 3\alpha = \cos(2\alpha + \alpha) = (\cos 2\alpha)(\cos \alpha) - (\sin 2\alpha)(\sin \alpha)$$

$$= (2 \cos^2 \alpha - 1)(\cos \alpha) - (2 \sin \alpha \cos \alpha)(\sin \alpha)$$

$$= (2 \cos^2 \alpha - 1)(\cos \alpha) - (2 \cos \alpha)(1 - \cos^2 \alpha)$$

$$= 4 \cos^2 \alpha - 3 \cos \alpha.$$

In considering whether a $60°$ angle is trisectable, let $\alpha = 20°$ and $x = \cos \alpha$. From Figure 1.4.16, we see that $\cos 60 = \cos 3\alpha = 1/2$. Then $4x^3 - 3x = 1/2$. This equation may be rewritten as $8x^3 - 6x - 1 = 0$. The question now becomes whether the roots of this equation are constructible. By the Rational Root Theorem, we know that if there are rational roots to this equation, they must come from the list $\{\pm 1, \pm 1/2, \pm 1/4, \pm 1/8\}$. Since none of these values satisfy the equation, there are no rational roots to the equation. Since any irrational roots would be cubic in nature (rather than a power of 2), by Theorem 1.4.2, $|\cos 3\alpha|$ is not constructible.

**URL
1.4.7**

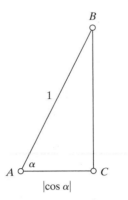

Figure 1.4.15 Constructible Angle

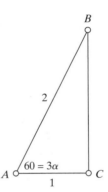

Figure 1.4.16 Trisecting an Angle

Summary

In the history of geometry, three straightedge and compass constructions captured the imagination and focused the attention of mathematicians from one generation to the next: Trisecting an angle; squaring the circle; and doubling the cube. The efforts of these geometers led to alternative solutions not restricted to the use of straightedge and compass and advanced mathematical understanding in other ways. The question of whether these constructions are possible using straightedge and compass was finally answered in the nineteenth century, ending their open status after more than 2000 years. Table 14.1 summarizes the definitions and theorems discused in this section.

Definitions	
1.4.1	A number *a* is *constructible* if one may construct a line segment with length $|a|$ in a finite number of steps from a segment of unit length using only a straightedge and compass.

Theorems	
1.4.1	If *a* and *b* are constructible real numbers, then so are a) $a + b$ b) $a - b$ c) ab d) a/b e) $\sqrt{a}$
1.4.2	The field of constructible real numbers consists of all real numbers obtained by taking square roots of positive rational numbers a finite number of times and applying a finite number of field operations.

Table 1.4.1 Summary, Section 1.4

URLs		Note: Begin each URL with the prefix http://
1.4.1	h	www-history.mcs.st-and.ac.uk/ history/Mathematicians/Hippocrates.html
1.4.2	h	www-history.mcs.st-and.ac.uk/ history/Mathematicians/Ahmes.html
1.4.3	h	www-history.mcs.st-and.ac.uk/ history/Mathematicians/Anaxagoras.html
1.4.4	h	www-history.mcs.st-and.ac.uk/ history/Mathematicians/Archimedes.html
1.4.5	h	www-history.mcs.st-and.ac.uk/ history/Mathematicians/Eratosthenes.html
1.4.6	h	www-history.mcs.st-and.ac.uk/ history/Mathematicians/Wantzel.html
1.4.7	h	library.advanced.org/10030/8trrt.htm

Table 1.4.2 Section 1.4 URLs (c = concept, h = history, s = software, d = data)

Exercises

1. The following angles may be trisected using only straightedge and compass. Demonstrate and justify trisections for the following angles:
   a) 90°
   b) 135°
   c) 45°
2. With respect to the Method of Hippocrates for trisecting an angle (see Figure 1.4.4), what about his approach violates the "straightedge and compass only" restriction of the original problem statement? Why does his approach work?

3. With respect to Archimedes' approach to squaring the circle, demonstrate a means for constructing a square with area equal to the right triangle ABC in Figure 1.4.6.

4. With respect to Figure 1.4.7, show that the edge of the outside square is equal to the edge of the shaded square times the square root of two.

5. With respect to Figure 1.4.12, use similar triangles to show that segment AE has length ab.

6. With respect to Figure 1.4.13, use similar triangles to show that segment AB has length a/b.

7. With respect to Figure 1.4.14, use similar triangles to show that segment DB has length $\sqrt{a}$.

8. Prove that it is impossible to square the circle using straightedge and compass.

9. Prove that it is impossible to double the cube using straightedge and compass.

Investigations

Origami Trisection of an Angle

One of the alternative methods developed to trisect acute angles uses the paper-folding approach of origami. The following demonstration shows one such method. Follow the directions to create an origami trisection. First, crease a square piece of paper to define an acute angle (see Figure 1.4.17).

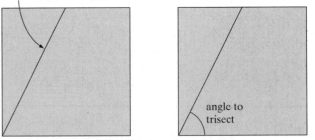

Figure 1.4.17 Angle to Trisect

When the trisection is completed, the paper will be creased as shown in Figure 1.4.18.

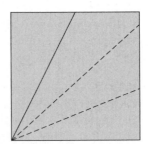

Figure 1.4.18 Results Sought

The approach involves finding three congruent triangles as shown in Figure 1.4.19.

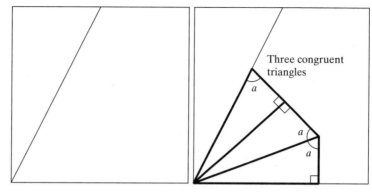

Figure 1.4.19 Approach

Step 1: Choose some height for the lower triangle, any height, and crease a horizontal line at this height (see Figure 1.4.20).

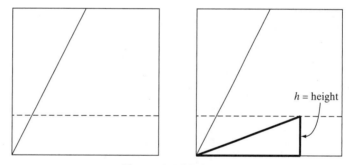

Figure 1.4.20 Step 1

We need to find a line of length $2h$ as shown in Figure 1.4.21.

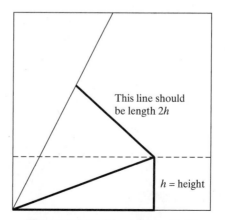

Figure 1.4.21 Line of Length $2h$

Step 2: Make a "marked ruler" in the side of the paper, by folding over the paper again as shown in Figure 1.4.22.

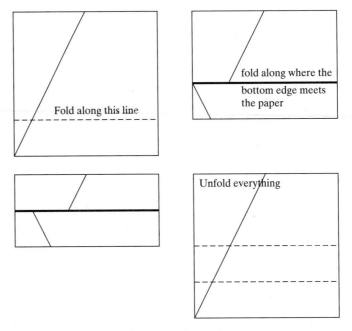

Figure 1.4.22 Ruler

Step 3: Now this "marked ruler" is used to find the bold line we needed (see Figure 1.4.23).

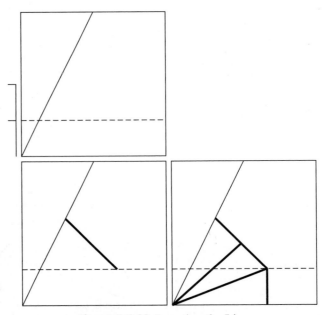

Figure 1.4.23 Locating the Line

Fold the paper so point *b* touches line *B*, and point *d* touches line *D*. Verify that the angle is trisected by measuring the three smaller angles with a protractor (see Figure 1.4.24).

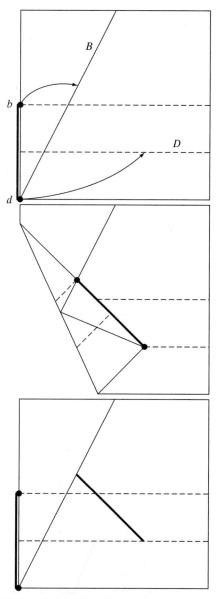

Figure 1.4.24 Folding the Paper: Part 1

From this point, how would you complete the trisection? Figure 1.4.25 illustrates some additional features of the figure.

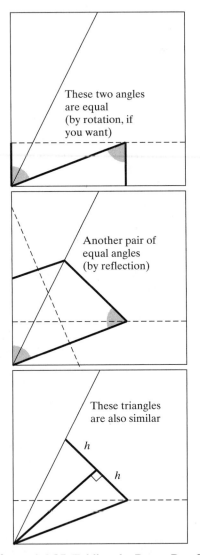

These two angles
are equal
(by rotation, if
you want)

Another pair of
equal angles
(by reflection)

These triangles
are also similar

h

h

Figure 1.4.25 Folding the Paper: Part 2

References and Suggested Reading

Armstrong, M.A. 1996. *Monetary History of Egypt: Ptolemaic Dynasty. Monetary History of the World.* Available on-line http://peicommerce.com/HISTORY/EUROPE/GREECE/EGYPT/PTOL-1.HTM

Bidwell, J.K. 1986. "A Babylonian geometrical algebra." *College Mathematical Journal,* 17 (1) 22–31.

Boyer, C.B. and U.C. Merzbach 1989. *A History of Mathematics,* 2nd ed. New York: John Wiley & Sons.

Bruins, E.M. 1981. Egyptian arithmetic. Janus 68 (1–3), 33–52.

Brundige, E.N. 1996. *The Library of Alexandria.* Perseus Project. Available on-line http://www.perseus.tufts.edu/GreekScience/Students/Ellen/Museum.html

Cajori, F. 1987. *A History of Elementary Mathematics.* London: Maximillian.

Dauben, J.W. 1985. *The History of Mathematics from Antiquity to the Present:* A Selective Bibliography. New York: Garland Publishing, Inc.

Fowler, D.H. 1987. *The Mathematics of Plato's Academy: A New Reconstruction.* Oxford: Clarendon Press.

Friberg, J. 1981. "Methods and traditions of Babylonian mathematics. II: An old Babylonian catalogue text with equations for squares and circles." *Journal of Cuneiform Studies,* 33 (1), 57–64.

Friberg, J. 1981. *Methods and traditions of Babylonian mathematics.* Plimpton 322, *Pythagorean triples, and the Babylonian triangle parameter equations.* Historia Mathematica 8, 277–318.

Gerdes, P. 1985. *Three alternate methods of obtaining the ancient Egyptian formula for the area of a circle.* Historia Mathematica 12 (3), 261–268.

Gillings, R.J. 1982. *Mathematics in the Time of the Pharaohs.* Cambridge, MA.

Heath, T.H. 1981. *A History of Greek Mathematics,* Vol. 1: *From Thales to Euclid.* New York: Dover.

Heath, T.H. 1981. *A History of Greek Mathematics,* Vol. 2: *From Aristarchus to Diophantus.* New York: Dover.

Interactive Models of Platonic and Archimedean Solids. Science U. Available on-line http://www.scienceu.com/geometry/facts/solids/handson.html

Joyce, D. 1999. *Euclid's Elements.* Available on-line at aleph0.clarku.edu/~djoyce/java/elements/toc.html

Lucas, J.R. 1961. "Minds, Machines and Gödel." *Philosophy, XXXVI,* 112–127. Available on-line http://users.ox.ac.uk/~jrlucas/mmg.html

Morrow, G.R. 1970. *Proclus; A Commentary on the First Book of Euclid's Elements.* [Translation] Princeton: Princeton University Press.

Netz, R. 1999. *The Shaping of Deduction in Greek Mathematics.* Cambridge, England: Cambridge University Press.

O'Connor, J.J. & E.F. Robertson, 1999. *The MacTutor History of Mathematics archive.* School of Mathematics and Statistics, University of St. Andrews, Scotland. Available on-line:

Ames. http://www-history.mcs.st-and.ac.uk/history/ Mathematicians/ Ames.html

Anaxagoras of Clazomenae. http://www-history.mcs.st-and.ac.uk/history/Mathematicians/Anaxagoras. html

Archimedes of Syracuse. http://www-history.mcs.st-and.ac.uk/history/ Mathematicians/ Archimedes.html

Aristotle. http://www-history.mcs.st-and.ac.uk/history/Mathematicians/ Aristotle.html

Babylonian and Egyptian mathematics. http://www-history.mcs.st-and.ac.uk/history/HistTopics/Babylonian_and_Egyptian.html

Chronology of the Greek period: 650 B.C.–500 A.D. http://www-history.mcs.st-and.ac.uk/history/Chronology/ChronologyA.html

Eratosthenes of Cyrene. http://www-history.mcs.st-and.ac.uk/history/Mathematicians/Eratosthenes.html

Euclid of Alexandria. http://www-history.mcs.st-and.ac.uk/history/Mathematicians/ Euclid.html

Eudoxus of Cnidus. http://www-history.mcs.st-and.ac.uk/history/Mathematicians/ Eudoxus.html

Godel. http://www-history.mcs.st-and.ac.uk/history/Mathematicians/ Godel.html

Hippocrates of Chios. http://www-history.mcs.st-and.ac.uk/history/Mathematicians/Hippocrates.html

Index of Greek mathematicians. http://www-history.mcs.st-and.ac.uk/history/ Indexes/ Greek_index.html

Plato. http://www-history.mcs.st-and.ac.uk/history/Mathematicians/Plato.html

Pythagoras of Samos. http://www-history.mcs.st-and.ac.uk/history/ Mathematicians/Pythagoras.html

Thales of Miletus. http://www-history.mcs.st-and.ac.uk/history/ Mathematicians/Thales.html

Wantzel. http://www-history.mcs.st-and.ac.uk/history/Mathematicians/ Wantzel.html

Rational Root Theorem. ThinkQuest Library. Available on-line at http://library. thinkquest.org/10030/8trrt.htm#top

Rees, C.S. 1981. "Egyptian fractions, Math." *Chronicle* 10 (1–2), 13–30.

Singmaster, D. 1985. "The legal values of pi, Math." *Intelligencer* 7 (2), 69–72.

Yaglom, I. 1995. "Number systems: Mayans, Romans, Babylonians–lend us your calculators," *Quantum* 5 (6), 23–27.

Chapter 2

Topics in Euclidean Geometry

URL
2.1.1

For over two thousand years, the study of geometry has been regarded as an essential element of a well-rounded education, introducing students to practical applications of mathematics, mathematical models, and the axiomatic method of proof. Euclid's *Elements* has provided a rich context for that undertaking for a hundred generations. Beginning in ancient Greece and continuing to this day, enlightened geometry teachers have always sought to develop in their students a deep understanding of both the content and the methods of geometry. In the tutorial tradition used extensively in the ancient and modern world prior to this century, teachers met regularly with small groups of students to explore new ideas, to formulate and test conjectures, to ask "why," and to present their findings as formal proofs.

During the twentieth century, education was reshaped in the image of industry. Large classes with periods of fixed length and syllabi that emphasize periodic testing of factual information and algorithms conspire to force geometric instruction into a mold it was never meant to occupy. As a consequence, many of today's high school students are asked to prove and apply theorems that they do not understand. No wonder geometry is the last mathematics course taken by most Americans.

URL
2.1.2

Fortunately, genuine reform is underway in geometric teaching and learning. The National Council of Teachers of Mathematics (2000) *Principles and Standards for School Mathematics* provides a conceptual roadmap for making the study of geometry more meaningful and rewarding for students. The section of that document dealing with geometry at the high school level states that

> By high school, students should have command of a broad range of geometric objects and properties. They should have had experience making and justifying simple conjectures about the relationships among those objects. During grades 9–12 the complexity of the relationships, whether presented statically or dynamically, must be extended and deepened. Using dynamic geometry software or physical models, students can quickly explore a range of examples. They can analyze what seems to change and what seems to remain invariant, and they can create conjectures about a given geometric situation. For example, students may notice that the diagonals of a parallelogram appear to meet at their respective midpoints. Many students may be content to stop at this point, convinced that their observation must generalize because it works for so many examples. However, effective teachers must challenge such assumptions.

They can take advantage of these opportunities to encourage students to develop deeper understandings by formulating verifiable conjectures, exploring possible explanations, and finally resolving them through counterexamples or proofs.

The importance of providing students with extensive informal geometric training during elementary and middle school cannot be overstated. Generally speaking, high-school students who have *informally* studied the concepts presented in a *formal* proposition are better able to appreciate the logical structure of the proposition's proof than students lacking this sort of preparation. Forced to master all aspects of the proof simultaneously and faced with a conceptual overload, these students frequently fail to grasp the purpose, method, and structure of the proof... that is, they "cannot see the forest for the trees." Over time, many of these students come to regard formal proofs as a bewildering array of statements and justifications rather than as a strategic deployment of related ideas.

The National Council of Teachers of Mathematics (1999) *Principles and Standards for School Mathematics–Working Draft* also states that

A critical element in the study of geometry for students in grades 9–12 is knowing how to judge, construct, and communicate proofs. Electronic technology enables students to explore geometric relationships dynamically, and offers students visual feedback and measurement as they investigate geometric situations. Secondary school teachers face an important challenge in balancing this use of technology for exploring ideas and developing conjectures with the use of deductive reasoning and counterexample in establishing or refuting the validity of such conjectures. Students should be able to articulate for themselves why particular conclusions about geometric objects or relationships among objects are logically sound. It is not critical that students master any particular format for presenting proofs (such as the two-column format). What is critical, however, is that students see the power of proof in establishing general claims (theorems) and that they are able to communicate their proofs effectively in writing.

URLs
2.1.3
&
2.1.4
The most popular of the dynamic geometry software packages indicated above is the *Geometers Sketchpad*. This software package provides convenient methods for constructing points, lines, arcs and circles. Triangles, polygons, and other objects created using these tools also may be measured, shaded, and manipulated in order to investigate their geometric features. In these Euclidean modeling environments, the most natural activities are constructing objects, investigating their features, formulating conjectures about their properties, testing the user's conjectures, and searching for the logical bases for relationships.

Euclid would rejoice at this development, for this sort of activity is the very essence of geometry. And just as constructing, investigating, conjecturing, and testing led to understanding and proof in his day, the same activities conducted in today's geometric modeling environments may be used to motivate and facilitate proof. To well-prepared students, proof is not distasteful or defeating; it is the ultimate and natural answer to the question "why?" In order to realize this goal, all students should approach the study of geometry through a gradual process beginning in elementary school with descriptive geometry and culminating in high school with the study of formal proof. The elements of that process and their natural sequencing may be summarized as follows:

Descriptive Geometry ▶ Geometric Constructions ▶ Observation ▶ Conjecture ▶ Testing ▶ Belief ▶ Informal Explanation ▶ Proof

This chapter illustrates various aspects of this sequence in the context of selected topics from Euclidean geometry through the ages.

2.1 Elementary Constructions

In this section, you will . . .

- **Investigate the purposes and methods of Euclidean constructions**

Learning to sketch natural and man-made objects is a challenging and intrinsically rewarding activity for students of all ages. Consequently, many students are genuinely interested in learning how to do elementary Euclidean constructions and are curious about their applications in mathematics, science, and the arts. By practicing elementary constructions, applying them in meaningful contexts, and developing fluency in the use of the language and tools used to describe and create geometric objects, students lay the foundation for systematic investigations of more abstract concepts and procedures.

In the *Elements,* geometric constructions are presented in the same manner as other propositions, justifying each step with axioms, common notions, definitions, and other propositions as necessary. These constructions permit the use of only two instruments, straightedge and compass. A straightedge is just that, a straight edge, and is to be used for one purpose only, drawing line segments between known points (see Figure 2.1.1). While rulers are often used as straightedges, the evenly spaced markings along their edges must never be used to measure distances or locate points. A compass (see Figure 2.1.2) is a device that may be used to draw an arc of constant radius from a fixed point, or center. Like rulers, some adjustable compasses are calibrated in a manner that may be used to determine the angular measure formed by the two arms of the compass. Any use of such rulings is not allowed in Euclidean constructions.

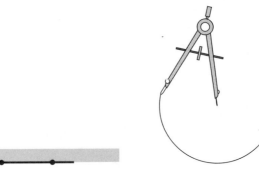

Figure 2.1.1 Straightedge   **Figure 2.1.2** Compass

One might argue that by imposing a "straightedge and compass only" approach to constructions, Euclid placed some important questions outside the scope of the geometric method (e.g., trisecting angles). On the other hand, because of this decision, students and teachers of geometry were forced to seek, and consequently found, a deeper understanding of geometry's most fundamental and powerful features than they might have otherwise. Perhaps for that reason, Euclid's approach to construction is still highly valued by educators.

Constructions as Instruction Sets

As a first step in learning Euclidean constructions, students should learn to read, follow, and create their own explicit instruction sets. Instruction sets should present an unambiguous sequence of actions using correct terminology and notation. For instance, the steps in constructing an equilateral triangle may be stated as shown in Table 2.1.1.

Step	Action
1	Using point A as the center and the length of segment AB as the radius, draw circle A.
2	Using point B as the center and the length of segment AB as the radius, draw circle B.
3	Mark one of the intersections of circle A and circle B as point C.
4	Connect point C to points A and B, forming triangle ABC.

Table 2.1.1 Instruction Set: Constructing an Equilateral Triangle

Constructions based on instruction sets should be practiced first using straightedge and compass then using dynamic geometric software, such as the *Geometers Sketchpad*. Using the *Geometers Sketchpad* and the instructions in Table 2.1.1, the equilateral triangle shown in Figure 2.1.3 is obtained.

**CD
2.1.1**

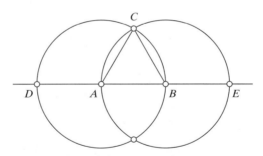

Figure 2.1.3 Constructing an Equilateral Triangle

As students master constructions based on instruction sets and begin to sense their underlying logic, they should be encouraged to explain why, for instance, point C is equidistant from points A and B and why triangle ABC is called equilateral in Figure 2.1.3. Using this approach, students gradually develop confidence in their ability to identify and to formulate conjectures and to offer reasonable explanations for geometric relationships. Repeated experiences and successes on tasks of this sort build student belief both in the validity of their findings and their ability as geometers. With this sort of background, students may approach the challenge of formal proof without having to simultaneously develop all the prerequisite geometric insights necessary to the task.

Constructions as Formal Propositions

**URL
2.1.5** The first proposition presented in Book I of Euclid's *Elements* illustrates and justifies the use of compass and straight edge in the construction of an equilateral triangle. The steps in the construction appear in the left column of Figure 2.1.4, with justifications in the right column. Justifications (see Table 2.1.2) include axioms (5 total), common notions (5 total), definitions and proven propositions (vary in number from book to book). For a listing of Euclid's axioms, common notions, and definitions and propositions for Book I, see Euclid's *Elements* on-line at David Joyce's WWW site.

Book I, Proposition 1 To construct an equilateral triangle on a given finite straight line.

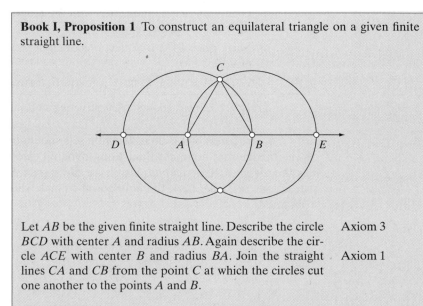

Let *AB* be the given finite straight line. Describe the circle *BCD* with center *A* and radius *AB*. Again describe the circle *ACE* with center *B* and radius *BA*. Join the straight lines *CA* and *CB* from the point *C* at which the circles cut one another to the points *A* and *B*.	Axiom 3 Axiom 1
Now, since the point *A* is the center of the circle *CDB*, therefore *AC* equals *AB*. Again, since the point *B* is the center of the circle *CAE*, therefore *BC* equals *BA*.	I Def. 15
But *AC* was proved equal to *AB*, therefore each of the straight lines *AC* and *BC* equals *AB*. And things which equal the same thing also equal one another, therefore *AC* also equals *BC*. Therefore the three straight lines *AC*, *AB*, and *BC* equal one another.	C.N. 1
Therefore the triangle *ABC* is equilateral, and it has been constructed on the given finite straight line *AB*.	I Def. 20

Figure 2.1.4 Book I, Proposition 1

**Example
2.1.1** To construct an equilateral triangle on a given finite straight line. Compare the steps listed in Proposition 1 to the instruction set in Table 2.1.1. How does Euclid's use of language differ from that used in Table 2.1.1?

Justification	Sample Notation
Axioms	Axiom 1
Common Notions	C.N. 2
Definitions	I Def. 20 (Book I, Definition 20)
Proven Propositions	IV Prop. 5 (Book IV, Proposition 5)

Table 2.1.2 Justification Notation

Example 2.1.2   To bisect a given rectilinear angle. One of the first Euclidean constructions taught in school is that of bisecting a given angle. Modern geometry texts typically present the construction as shown in Table 2.1.3.

Step	Action
1	Draw arc *DE* with center *A*.
2	Using the same radius as arc *DE*, draw arcs *OQ* with center *E* and *KM* with center *D*.
3	Mark the intersection of arcs *OQ* and *KM* as point *H*.
4	Draw ray *AH*.

Table 2.1.3 Instruction Set: Bisecting an Angle

Following the instruction set in Table 2.1.3 and using the *Geometers Sketchpad*, Figure 2.1.5 is obtained.

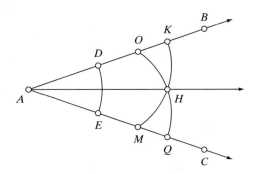

Figure 2.1.5 Bisecting an Angle

The sequence of steps in Table 2.1.3 is somewhat different than that presented in Euclid's version (Proposition 9) of this construction (see Figure 2.1.6). Represent Euclid's approach in an instruction set like that shown in Table 2.1.3 and demonstrate the construction using straightedge and compass and the *Geometers Sketchpad*.

Proposition 9 To bisect a given rectilinear angle.

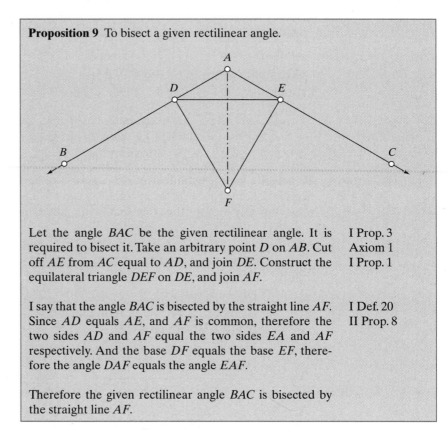

Let the angle *BAC* be the given rectilinear angle. It is required to bisect it. Take an arbitrary point *D* on *AB*. Cut off *AE* from *AC* equal to *AD*, and join *DE*. Construct the equilateral triangle *DEF* on *DE*, and join *AF*.

I Prop. 3
Axiom 1
I Prop. 1

I say that the angle *BAC* is bisected by the straight line *AF*. Since *AD* equals *AE*, and *AF* is common, therefore the two sides *AD* and *AF* equal the two sides *EA* and *AF* respectively. And the base *DF* equals the base *EF*, therefore the angle *DAF* equals the angle *EAF*.

I Def. 20
II Prop. 8

Therefore the given rectilinear angle *BAC* is bisected by the straight line *AF*.

Figure 2.1.6 Book I, Proposition 9

Example 2.1.3

To bisect a given finite straight line. Figure 2.1.7 demonstrates Euclid's method for bisecting a line segment (Proposition 10). As in Proposition 9 (see Figure 2.1.6), an equilateral triangle is created as a strategic element in the construction. The proposition also uses Proposition 4 as justification: If two triangles have two sides equal to two sides respectively, and have the angles contained by the equal straight lines equal, then they also have the base equal to the base, the triangle equals the triangle, and the remaining angles equal the remaining angles respectively, namely those opposite the equal sides. In modern geometry texts, this proposition is often identified as the Side-Angle-Side Theorem.

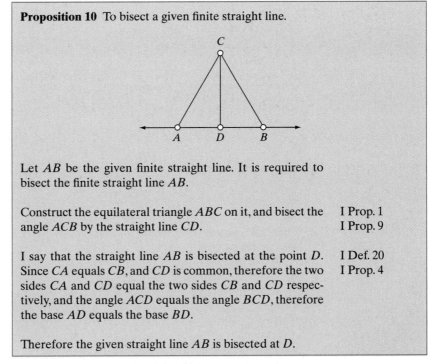

Proposition 10 To bisect a given finite straight line.

Let *AB* be the given finite straight line. It is required to bisect the finite straight line *AB*.

Construct the equilateral triangle *ABC* on it, and bisect the   I Prop. 1
angle *ACB* by the straight line *CD*.   I Prop. 9

I say that the straight line *AB* is bisected at the point *D*.   I Def. 20
Since *CA* equals *CB*, and *CD* is common, therefore the two   I Prop. 4
sides *CA* and *CD* equal the two sides *CB* and *CD* respectively, and the angle *ACD* equals the angle *BCD*, therefore the base *AD* equals the base *BD*.

Therefore the given straight line *AB* is bisected at *D*.

Figure 2.1.7 Book I, Proposition 10

After reviewing Euclid's approach in Proposition 10, write a sequence of steps for bisecting a segment that does not involve creating an equilateral triangle. Prove your method using justifications acceptable to Euclid and demonstrate the construction using straightedge and compass and the *Geometers Sketchpad*.

Example 2.1.4   To draw a straight line at right angles to a given straight line from a given point on it (see Figure 2.1.8).

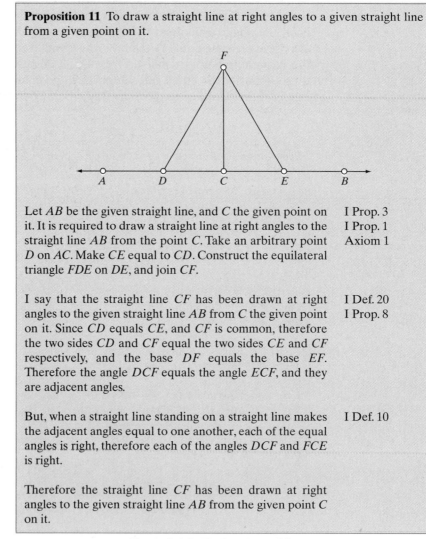

Proposition 11 To draw a straight line at right angles to a given straight line from a given point on it.

Let *AB* be the given straight line, and *C* the given point on it. It is required to draw a straight line at right angles to the straight line *AB* from the point *C*. Take an arbitrary point *D* on *AC*. Make *CE* equal to *CD*. Construct the equilateral triangle *FDE* on *DE*, and join *CF*.	I Prop. 3 I Prop. 1 Axiom 1
I say that the straight line *CF* has been drawn at right angles to the given straight line *AB* from *C* the given point on it. Since *CD* equals *CE*, and *CF* is common, therefore the two sides *CD* and *CF* equal the two sides *CE* and *CF* respectively, and the base *DF* equals the base *EF*. Therefore the angle *DCF* equals the angle *ECF*, and they are adjacent angles.	I Def. 20 I Prop. 8
But, when a straight line standing on a straight line makes the adjacent angles equal to one another, each of the equal angles is right, therefore each of the angles *DCF* and *FCE* is right.	I Def. 10
Therefore the straight line *CF* has been drawn at right angles to the given straight line *AB* from the given point *C* on it.	

Figure 2.1.8 Book I, Proposition 11

After reviewing Euclid's approach in Proposition 11, write a sequence of steps for constructing the perpendicular to line *AB* at point *C* that does not involve creating an equilateral triangle. Prove your method using justifications acceptable to Euclid and demonstrate the construction using straightedge and compass and the *Geometers Sketchpad*.

Example 2.1.5   To draw a straight line perpendicular to a given infinite straight line from a given point not on it (see Figure 2.1.9).

Proposition 12 To draw a straight line perpendicular to a given infinite straight line from a given point not on it.

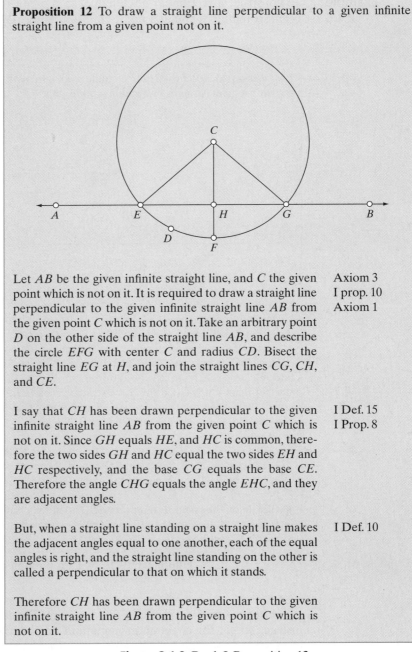

Let *AB* be the given infinite straight line, and *C* the given point which is not on it. It is required to draw a straight line perpendicular to the given infinite straight line *AB* from the given point *C* which is not on it. Take an arbitrary point *D* on the other side of the straight line *AB*, and describe the circle *EFG* with center *C* and radius *CD*. Bisect the straight line *EG* at *H*, and join the straight lines *CG*, *CH*, and *CE*.	Axiom 3 I prop. 10 Axiom 1
I say that *CH* has been drawn perpendicular to the given infinite straight line *AB* from the given point *C* which is not on it. Since *GH* equals *HE*, and *HC* is common, therefore the two sides *GH* and *HC* equal the two sides *EH* and *HC* respectively, and the base *CG* equals the base *CE*. Therefore the angle *CHG* equals the angle *EHC*, and they are adjacent angles.	I Def. 15 I Prop. 8
But, when a straight line standing on a straight line makes the adjacent angles equal to one another, each of the equal angles is right, and the straight line standing on the other is called a perpendicular to that on which it stands.	I Def. 10
Therefore *CH* has been drawn perpendicular to the given infinite straight line *AB* from the given point *C* which is not on it.	

Figure 2.1.9 Book I, Proposition 12

After reviewing Euclid's approach in Proposition 12, write a sequence of steps for constructing the perpendicular to line *AB* through point *C* that does not involve creating an equilateral triangle. Prove your method using justifications acceptable to Euclid and demonstrate the construction using straightedge and compass and the *Geometers Sketchpad*.

**Example
2.1.6** To construct a rectilinear angle equal to a given rectilinear angle on a given straight line and at a point on it (see Figure 2.1.10).

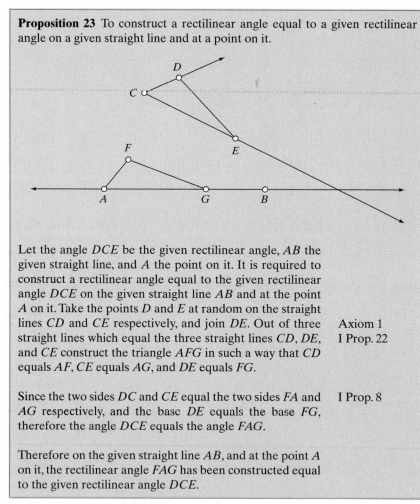

Proposition 23 To construct a rectilinear angle equal to a given rectilinear angle on a given straight line and at a point on it.

Let the angle *DCE* be the given rectilinear angle, *AB* the given straight line, and *A* the point on it. It is required to construct a rectilinear angle equal to the given rectilinear angle *DCE* on the given straight line *AB* and at the point *A* on it. Take the points *D* and *E* at random on the straight lines *CD* and *CE* respectively, and join *DE*. Out of three straight lines which equal the three straight lines *CD*, *DE*, and *CE* construct the triangle *AFG* in such a way that *CD* equals *AF*, *CE* equals *AG*, and *DE* equals *FG*.

    Axiom 1
    I Prop. 22

Since the two sides *DC* and *CE* equal the two sides *FA* and *AG* respectively, and the base *DE* equals the base *FG*, therefore the angle *DCE* equals the angle *FAG*.

    I Prop. 8

Therefore on the given straight line *AB*, and at the point *A* on it, the rectilinear angle *FAG* has been constructed equal to the given rectilinear angle *DCE*.

Figure 2.1.10 Book I, Proposition 23

After reviewing Euclid's approach in Proposition 23, write an instruction set for copying an angle and demonstrate the construction using straightedge and compass and the *Geometers Sketchpad*.

Example 2.1.7   To draw a straight line through a given point parallel to a given straight line (see Figure 2.1.11).

Proposition 31 To draw a straight line through a given point parallel to a given straight line.

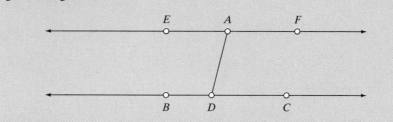

Let *A* be the given point, and *BC* the given straight line. It is required to draw a straight line through the point *A* parallel to the straight line *BC*.

Take a point *D* at random on *BC*. Join *AD*. Construct the angle *DAE* equal to the angle *ADC* on the straight line *DA* and at the point *A* on it. Produce the straight line *AF* in a straight line with *EA*.

Axiom 1
I Prop. 23
Axiom 2

Since the straight line *AD* falling on the two straight lines *BC* and *EF* makes the alternate angles *EAD* and *ADC* equal to one another, therefore *EAF* is parallel to *BC*.

I Prop. 27

Therefore the straight line *EAF* has been drawn through the given point *A* parallel to the given straight line *BC*.

Figure 2.1.11 Book I, Proposition 31

After reviewing Euclid's approach in Proposition 31, write an instruction set for drawing a straight line through a given point parallel to a given straight line and demonstrate the construction using straight-edge and compass and the *Geometers Sketchpad*.

Example 2.1.8 To describe a square on a given straight line (see Figure 2.1.12).

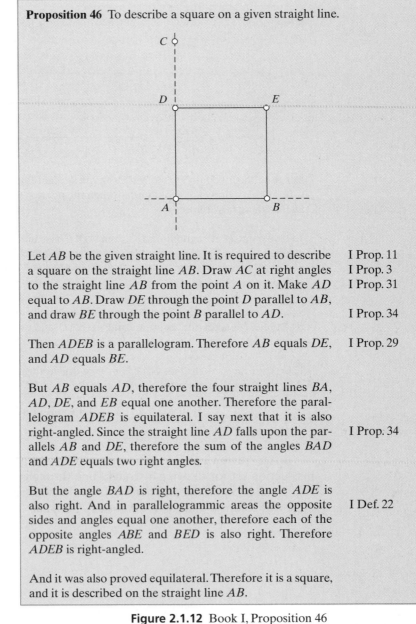

Proposition 46 To describe a square on a given straight line.

Let *AB* be the given straight line. It is required to describe a square on the straight line *AB*. Draw *AC* at right angles to the straight line *AB* from the point *A* on it. Make *AD* equal to *AB*. Draw *DE* through the point *D* parallel to *AB*, and draw *BE* through the point *B* parallel to *AD*.   I Prop. 11 / I Prop. 3 / I Prop. 31 / I Prop. 34

Then *ADEB* is a parallelogram. Therefore *AB* equals *DE*, and *AD* equals *BE*.   I Prop. 29

But *AB* equals *AD*, therefore the four straight lines *BA*, *AD*, *DE*, and *EB* equal one another. Therefore the parallelogram *ADEB* is equilateral. I say next that it is also right-angled. Since the straight line *AD* falls upon the parallels *AB* and *DE*, therefore the sum of the angles *BAD* and *ADE* equals two right angles.   I Prop. 34

But the angle *BAD* is right, therefore the angle *ADE* is also right. And in parallelogrammic areas the opposite sides and angles equal one another, therefore each of the opposite angles *ABE* and *BED* is also right. Therefore *ADEB* is right-angled.   I Def. 22

And it was also proved equilateral. Therefore it is a square, and it is described on the straight line *AB*.

Figure 2.1.12 Book I, Proposition 46

After reviewing Euclid's approach in Proposition 46, write an instruction set for describing a square on a given straight line and demonstrate the construction using straightedge and compass and the *Geometers Sketchpad*.

Summary

Euclidean constructions are more than instruction sets on how to create geometric objects with well-defined features. Viewed as formal propositions, they are also explanations for why a given set of actions is guaranteed to result in a particular object. Consequently, students should do more than memorize instruction sets. They should learn to recognize the logic behind constructions and develop skill in the use of straightedge and compass and dynamic modeling tools such as the *Geometers Sketchpad*. With this fundamental grounding, they may begin exploring geometric relationships between objects.

URLs		Note: Begin each URL with the prefix http://
2.1.1	c	aleph0.clarku.edu/~djoyce/java/elements/ elements.html
2.1.2	c	standards-e.nctm.org/1.0/normal/standards/ standardsFS.html
2.1.3	h	www.keypress.com/sketchpad/index.html
2.1.4	h	www-cabri.imag.fr/index-e.html
2.1.5	h	aleph0.clarku.edu/~djoyce/java/elements/ bookI/propI1.html

Table 2.1.4 Section 2.1 URLs (c = concept, h = history, s = software, d = data)

Exercises

1. After reviewing Euclid's approach in each of the following propositions, write an instruction set for the construction. Then, demonstrate the construction using both straightedge and compass and the *Geometers Sketchpad*.
   a) Proposition 9
   b) Proposition 10
   c) Proposition 11
   d) Proposition 12
   e) Proposition 23
   f) Proposition 31
   g) Proposition 46
2. Write an instruction set for constructing the bisectors of each angle of a triangle. Then, demonstrate the construction using both straightedge and compass and the *Geometers Sketchpad*.
3. Write an instruction set for constructing the medians to each side of a triangle. Then, demonstrate the construction using both straightedge and compass and the *Geometers Sketchpad*.
4. Write an instruction set for constructing the perpendicular bisectors of each side of a triangle. Then, demonstrate the construction using both straightedge and compass and the *Geometers Sketchpad*.

5. Write an instruction set for constructing the altitudes to each side of a triangle. Then, demonstrate the construction using both straightedge and compass and the *Geometers Sketchpad.*

6. Each of the following constructions is a Euclidean proposition. Search Euclid's *Elements* and identify each proposition by book and number, then briefly summarize any strategic or logical differences between your construction and Euclid's.

   a) To construct a parallelogram having a given interior angle. *prop 31*

   b) To inscribe a square in a given circle.

   c) To find a mean proportional to two given straight lines.

2.2   Exploring Relationships Between Objects

In this section, you will . . .

- **Focus on *Observation, Conjecture, Testing,* and *Belief* in the Sequence: Descriptive Geometry ► Geometric Constructions ► Observation ► Conjecture ► Testing ► Belief ► Informal Explanation ► Proof**

Constructions use a straightforward cause-and-effect process to create objects with specific geometric features. Before beginning, the student already knows that certain features will be evident when the construction is finished. In that regard, constructions are like thoroughly explained recipes. It is an entirely different matter to examine an object or objects in search of features and/or relationships not anticipated in their construction. For instance, while Proposition 46 gives complete instructions on how to create a square, it says nothing about additional features and relationships that might emerge when the squares' diagonals are drawn. Features of this sort must be discovered by observation and validated by logic.

Systematic observation is more like a scientific investigation than a recipe, i.e., you don't know ahead of time what you will discover. To complicate matters, many students have little knowledge or experience in how to conduct such investigations. This section examines three elements of that process, observing, conjecturing, testing and believing, and illustrates the value of dynamic geometry software in such investigations.

Example 2.2.1   The perpendicular bisectors of the sides of a triangle. To illustrate the role of dynamic software in facilitating observation, consider the instruction set in Table 2.2.1. Assume that step 3 is a well-defined operation.

Step	Action
1	Draw three noncollinear points A, B, and C.
2	Connect A, B, and C with line segments, forming triangle ABC.
3	Construct the perpendicular bisectors of segments AB, BC, and CA.

Table 2.2.1 Instruction Set: Perpendicular Bisectors of the Sides of a Triangle

This instruction set is guaranteed to produce the perpendicular bisectors of the sides of any triangle. No other features or relationships are suggested by the instruction set. It is only when the construction is completed that additional features become apparent. Using the *Geometers Sketchpad*, a dynamic model of the construction was created and used to produce the objects in Figure 2.2.1. Each object shows all three perpendicular bisectors intersecting in a common point. Moving the vertices about in the dynamic model produces similar results. In Case 1, the intersection is inside the triangle. In Case 2, the intersection is on the triangle. And in Case 3 the intersection is outside the triangle. These are *observations*.

**CD
2.2.1**

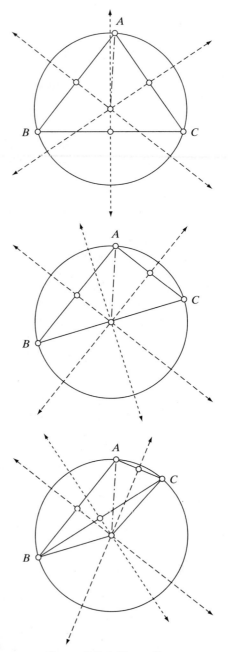

Figure 2.2.1 Three Cases

Because the perpendicular bisectors always seem to intersect in a common point, a tentative generalization seems in order: The perpendicular bisectors of the sides of a triangle intersect in a common point. This is a *conjecture*. A conjecture is an unproven assertion similar to a scientific hypothesis.

In order to *test* the conjecture "The perpendicular bisectors of the sides of a triangle intersect in a common point," it is important to understand what it would take to validate the hypothesis and what it would take to invalidate it. Since the hypothesis is about *all* triangles, validating the hypothesis must necessarily involve demonstrating that it is true for all triangles. Showing it is true for *some* triangles is not the same as showing that it is true for all triangles. Consequently, one must either demonstrate the truth of the hypothesis for every conceivable triangle, or prove it is true for all triangles using a logical argument. Since no exhaustive demonstration is feasible, the only way to validate the hypothesis is by formal proof. On the other hand, it would only take a single counterexample to invalidate the hypothesis because that single counterexample would prevent the hypothesis from being true for all triangles. What would a counterexample "look like?" It would be a triangle in which the three perpendicular bisectors fail to intersect at a common point. Using the *Sketchpad* model of the triangle, one may quickly satisfy oneself that no counterexamples seem likely. This leads to a well-informed *belief* that the conjecture is true.

While informed belief of this sort does not constitute proof, it is an essential prerequisite to proof. Systematic observation and reflection are the roots of insight and discovery in all of the sciences, mathematics included. Students who have thoroughly acquainted themselves with the conditions of a proposition through systematic observation are ready to ask "Why?"

Theorem 2.2.1 The perpendicular bisectors of the sides of every triangle intersect in a common point equidistant from the vertices of the triangle.

A proof of this theorem is left as an exercise.

Definition 2.2.1 The point of intersection of the perpendicular bisectors of the sides of a triangle is called the *circumcircle*.

Euclid introduces the circumcenter and circumcircle in Book IV Proposition 4 of the *Elements*. Further observation motivates the following questions about the circumcenter.

- Is it possible for the circumcenter to lie on a vertex of the triangle? If so, under what circumstances? If not, why not?
- Is it possible for the circumcenter to lie on a side of the triangle? If so, under what circumstances? If not, why not?
- If the triangle is equilateral, where is the circumcenter?

A similar approach may also be applied in an investigation of the features of other constructions. The following examples illustrate that approach.

Theorem 2.2.2 The bisectors of the angles of a triangle intersect at a common point equidistant from the sides of the triangle.

A proof of this theorem is left as an exercise.

Definition
2.2.2
The point of intersection of the angle bisectors of a triangle is called the *incenter.*

Example
2.2.2
Using the *Geometers Sketchpad,* construct the bisectors of the angles of a triangle as shown in Figure 2.2.2.

CD
2.2.2

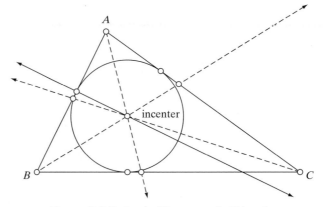

Figure 2.2.2 Angle Bisectors of a Triangle

Using the *Geometers Sketchpad's* (GSP) arrow icon, drag the vertices around. Based on your observations, answer the following questions.

- Is it possible for the incenter to lie on a vertex of the triangle? If so, under what circumstances? If not, why not?
- Is it possible for the incenter to lie on a side of the triangle? If so, under what circumstances? If not, why not?
- Is it possible for the incenter to lie outside of the triangle? If so, under what circumstances? If not, why not?
- State a conjecture concerning possible locations of the incenter. To prove your conjecture false, what sort of counter example would be necessary?
- If the triangle is equilateral, where is the incenter?
- Is it possible for the circumcenter and incenter to coincide? If so, under what circumstances? If not, why not?

Theorem
2.2.3
The altitudes of a triangle intersect at a point.

A proof of this theorem is left as an exercise.

Definition
2.2.3
The point of intersection of the altitudes of a triangle is called the *orthocenter.*

Example
2.2.3
Using the *Geometers Sketchpad,* construct the altitudes to the sides of a triangle as shown in Figure 2.2.3. Note that these altitudes intersect at a point called the orthocenter.

Using the GSP's arrow icon, drag the vertices around. Based on your observations, answer the following questions.

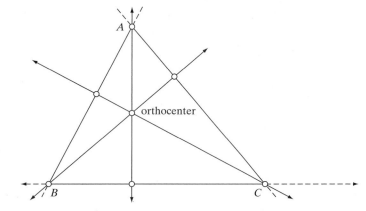

Figure 2.2.3 Orthocenter of a Triangle

- Is it possible for the orthocenter to lie on a vertex of the triangle? If so, under what circumstances? If not, why not?
- Is it possible for the orthocenter to lie on a side of the triangle? If so, under what circumstances? If not, why not?
- Is it possible for the orthocenter to lie outside of the triangle? If so, under what circumstances? If not, why not?
- State a conjecture concerning possible locations of the orthocenter. To prove your conjecture false, what sort of counter example would be necessary?
- If the triangle is equilateral, where is the orthocenter?

**Theorem
2.2.4**

The medians of a triangle intersect at a common point.

A proof of this theorem is left as an exercise.

**Definition
2.2.4**

The point of intersection of the medians of a triangle is called the *centroid*.

**Example
2.2.4**

Using the *Geometers Sketchpad,* construct the medians to the sides of a triangle as shown in Figure 2.2.4. Note that these medians intersect at a single point. This point is called the centroid.

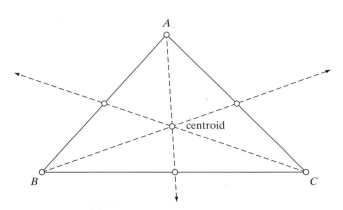

Figure 2.2.4 Centroid of a Triangle

Using the GSP's arrow icon, drag the vertices around. Based on your observations, answer the following questions.

- Is it possible for the centroid to lie on a vertex of the triangle? If so, under what circumstances? If not, why not?
- Is it possible for the centroid to lie on a side of the triangle? If so, under what circumstances? If not, why not?
- Is it possible for the centroid to lie outside of the triangle? If so, under what circumstances? If not, why not?
- State a conjecture concerning possible locations of the centroid. To prove your conjecture false, what sort of counter example would be necessary?
- Construct three triangle interiors, each of which includes point *A, B,* or *C,* has one vertex at the centroid, and its third vertex at a midpoint of a side adjacent to the selected vertex of the original triangle. Measure and compare the areas of these triangles.
- Using the GSP's arrow icon, drag the vertices around. State a conjecture based on your observations. To prove your conjecture false, what sort of counter example would be necessary?
- Why do you think the centroid is called the *center of gravity* of the triangle?

**Definition
2.2.5**

**URL
2.2.1**

The midpoints of the sides of a triangle, the points of intersection of the altitudes and the sides, and the midpoints of the segments joining the orthocenter and the vertices of a triangle all lie on the *nine-point circle*. The nine-point circle was discovered by Karl Wilhelm Feurbach (1800–1834).

**Example
2.2.5**

Using the GSP's arrow icon, drag the vertices around (see Figure 2.2.5). Based on your observations, answer the following questions.

**CD
2.2.5**

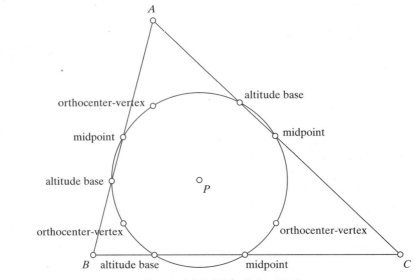

Figure 2.2.5 Nine-Point Circle

- Is it possible to position the vertices so that each side is tangent to the circle? If so, under what circumstances does this occur? If not, why not?
- When the triangle is isosceles, what happens to the nine points?

By adding additional lines, measuring segments, and computing ratios of segments (see Figure 2.2.6), the investigation may be extended to address new questions.

CD
2.2.6

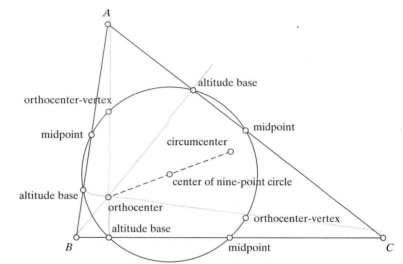

Figure 2.2.6 Adding Lines and Taking Measurements

- State a conjecture describing the relative positions of the orthocenter, circumcenter, and center of the nine-point circle. To prove your conjecture false, what sort of counterexample would be necessary?
- Is it possible to position the vertices so that the orthocenter, circumcenter, and center of the nine-point circle coincide? If so, under what circumstances? If not, why not?
- State a conjecture concerning the radius of the circumcircle of the triangle and the radius of the nine-point circle. To prove your conjecture false, what sort of counterexample would be necessary?
- State a conjecture describing the manner in which the nine-point circle divides segments drawn from the orthocenter to points on the circumcircle. To prove your conjecture false, what sort of counterexample would be necessary?

Learning to Use the *Geometers Sketchpad*

URL
2.2.2

In order to apply the ideas presented in this section, considerable skill is needed in the use of the geometry software, the *Geometers Sketchpad*. A free, 40 page workshop guide for the MS Windows and Macintosh versions of the *Geometers Sketchpad* is available on-line in PDF format. Working through this workshop guide takes a few hours but is well worth the time. Whether working through the workshop guide or the exercises and investigations in this text, most students find that working together in groups of two or three enhances their learning and reduces frustration and wasted time. Working together also encourages the development of communication skills and productive collaborations. The following example illustrates the use of the *Geometers Sketchpad* to identify relationships between angles, segments, and areas.

Example 2.2.6 In parallelogram $ABDC$ (Figure 2.2.7), point E is the midpoint of side CD. Segment BE and diagonal AD intersect at point F. Search for relationships involving the figure's angles, segments, and polygonal areas.

CD 2.2.7

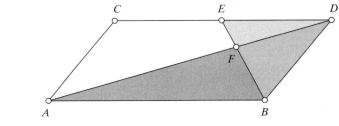

Figure 2.2.7 Divided Parallelogram

In the *Sketchpad* model of this figure, points A, B, and C may be moved independently of one another to reshape the parallelogram (see Figure 2.2.8). The other points in the figure are dependent on points A, B, and C. Each modification produces a new parallelogram.

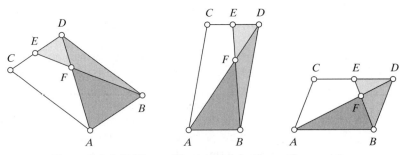

Figure 2.2.8 Different Versions of the Same Construction

By examining different versions of the same construction, many relationships between the objects' angles, segments, polygonal areas, and so on are easily recognized using the *Sketchpad's* Measure and Calculate tools. For instance, Table 2.2.2 shows that, regardless of the specific areas found in the four shaded areas of the parallelogram, the relationships between the areas remain the same. Those relationships are expressed in the following conjectures:

1. The area of triangle DFE is half that of triangle DFB.
2. The area of triangle DFB is half that of triangle FBA.
3. The area of triangle DFE is one-fifth that of quadrilateral $CEFA$.

Version 1	Version 2	Version 3
Area $DFE = 0.3.0$ cm^2	Area $DFE = 0.3$ cm^2	Area $DFE = 0.2$ cm^2
Area $DFB = 0.7.0$ cm^2	Area $DFB = 0.6$ cm^2	Area $DFB = 0.5$ cm^2
Area $FBA = 1.4.0$ cm^2	Area $FBA = 1.2$ cm^2	Area $FBA = 0.9$ cm^2
Area $CEFA = 1.7.0$ cm^2	Area $CEFA = 1.6$ cm^2	Area $CEFA = 1.2$ cm^2
$\dfrac{(\text{Area } DFE)}{(\text{Area } DFB)} = 0.500$	$\dfrac{(\text{Area } DFE)}{(\text{Area } DFB)} = 0.500$	$\dfrac{(\text{Area } DFE)}{(\text{Area } DFB)} = 0.500$
$\dfrac{(\text{Area } DFB)}{(\text{Area } FBA)} = 0.500$	$\dfrac{(\text{Area } DFB)}{(\text{Area } FBA)} = 0.500$	$\dfrac{(\text{Area } DFB)}{(\text{Area } FBA)} = 0.500$
$\dfrac{(\text{Area } DFE)}{(\text{Area } CEFA)} = 0.200$	$\dfrac{(\text{Area } DFE)}{(\text{Area } CEFA)} = 0.200$	$\dfrac{(\text{Area } DFE)}{(\text{Area } CEFA)} = 0.200$

Table 2.2.2 Area Relationships

By comparing the logical relationships among these conjectures, additional conjectures may be made, such as ...

4. The area of triangle *DFE* is one-fourth that of triangle *FBA*.
5. The area of triangle *FBA* is four-fifths that of quadrilateral *CEFA*.

The power of this approach is immediately evident and should be experienced first-hand to appreciate the satisfaction that comes with discovery. There are many other interesting relationships implicit in the construction of this divided parallelogram. Finding them is left as an exercise. Becoming expert in this approach is left as a life-long journey.

Summary

In learning to think like a geometer, students must develop observational and thinking skills that facilitate investigation of geometric objects and their relationships. In learning to formulate and test their own conjectures, students develop insights and thinking skills that are essential elements of mathematical inquiry and proof. They also develop fluency in the use of mathematical language, notation, and figures and confidence in their own ability to participate in mathematical dialogues. While modern technologies like the *Geometers Sketchpad* provide powerful, user-friendly modeling environments in which to conduct geometric investigations, they also impose additional learning requirements on both students and teachers. In spite of this, the long-term benefits associated with the use of geometry technologies greatly outweigh the temporary and/or occasional liabilities that they impose.

Table 2.2.3 summarizes the definitions discussed in this section.

Definitions	
2.2.1	The perpendicular bisectors of the sides of every triangle intersect in a common point called the *circumcenter*. Because this point is equidistant from the vertices of the triangle, it is also the center of the circle inscribing the triangle, called the *circumcircle*.
2.2.2	The bisectors of the angles of a triangle intersect at a point called the *incenter*.
2.2.3	The altitudes of a triangle intersect at a point called the *orthocenter*.
2.2.4	The medians of a triangle intersect at a point called the *centroid*.
2.2.5	The midpoints of the sides of a triangle, the points of intersection of the altitudes and the sides, and the midpoints of the segments joining the orthocenter and the vertices of a triangle all lie on the *nine-point circle*.
Theorems	
2.2.1	The perpendicular bisectors of the sides of every triangle intersect in a common point equidistant from the vertices of the triangle.
2.2.2	The bisectors of the angles of a triangle intersect at a common point equidistant from the sides of the triangle.
2.2.3	The altitudes of a triangle intersect at a point.
2.2.4	The medians of a triangle intersect at a common point.

Table 2.2.3 Summary, Section 2.2

URLs		Note: Begin each URL with the prefix http://
2.2.1	h	www-history.mcs.st-and.ac.uk/~history/ Mathematicians/Feuerbach.html
2.2.2	s	www.keypress.com/sketchpad/sketchdemo.html

Table 2.2.4 Section 2.2 URLs (c = concept, h = history, s = software, d = data)

Exercises

1. Extend the investigation begun in Example 2.2.1 by answering the following questions:
   a) Is it possible for the circumcenter to lie on a vertex of the triangle? If so, under what circumstances? If not, why not?
   b) Is it possible for the circumcenter to lie on a side of the triangle? If so, under what circumstances? If not, why not?
   c) If the triangle is equilateral, where is the circumcenter?
2. Answer the questions posed in each of the following examples.

a) 2.2.2
b) 2.2.3
c) 2.2.4
d) 2.2.5

3. Continue the investigation begun in Example 2.2.6, stating at least three additional conjectures.

4. Figure 2.2.6 suggests that the center of the nine-point circle might be the midpoint of the segment joining the orthocenter and circumcenter. Using the *Geometers Sketchpad,* construct a model with which to test this conjecture. Write an instruction set for this construction, then explain what it would take to validate this conjecture and what it would take to invalidate it. Discuss your findings and conclusion.

5. For which triangles does the circumcenter coincide with the incenter? Justify your answer.

6. Construct the following using both straightedge and compass.
   a) circumcenter
   b) incenter
   c) orthocenter
   d) centroid
   e) nine-point circle

7. The Euler line of a triangle contains the circumcenter, centroid, and orthocenter. Construct the Euler line for a triangle and formulate a conjecture regarding the relative positions and separations of these points.

8. Prove theorems
   a) 2.2.1
   b) 2.2.2
   c) 2.2.3
   d) 2.2.4

Investigations

Secant and Tangent Relationship
Tool(s) Geometers Sketchpad   *Data File(s)* cd2_2_8.gsp

Focus
Investigate the relationship between the lengths of the segments drawn from an exterior point on a secant line to the points of the intersection of the secant with a circle and the length of the tangent from the exterior point to the circle.

Tasks
1. Using the GSP's arrow icon, drag the vertices around and observe the measurements and calculations. State a conjecture based on your observations. To prove your conjecture false, what sort of counter example would be necessary?

2. Find two similar triangles in the figure on the next page and determine the ratio of their corresponding sides. Construct their polygon interiors and determine the ratio of their areas. How is the ratio of their areas related to the ratio of their sides?

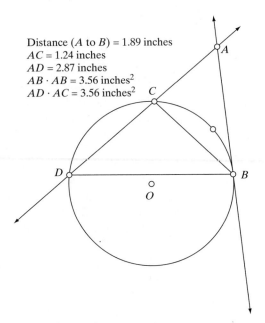

Distance $(A$ to $B) = 1.89$ inches
$AC = 1.24$ inches
$AD = 2.87$ inches
$AB \cdot AB = 3.56$ inches^2
$AD \cdot AC = 3.56$ inches^2

Segment Ratio
Tool(s) Geometers Sketchpad   *Data File(s)* cd2_2_9.gsp

Focus
A segment may be divided into 2 unequal but related parts.

Tasks
1. Using the GSP's arrow icon, drag points E and D around and observe the measurements and calculations. What do you notice about the ratios AE/CD and AB/BC? State a conjecture based on your observations. To prove your conjecture false, what sort of counter example would be necessary?
2. Write a paragraph discussing the geometric basis for this procedure.

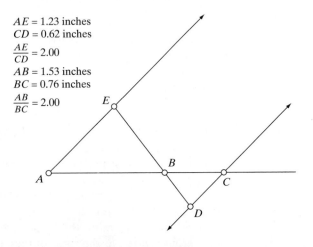

$AE = 1.23$ inches
$CD = 0.62$ inches
$\dfrac{AE}{CD} = 2.00$
$AB = 1.53$ inches
$BC = 0.76$ inches
$\dfrac{AB}{BC} = 2.00$

Miquel Point
Tool(s) Geometers Sketchpad   *Data File(s)* cd2_2_10.gsp

Focus
Any three noncollinear points determine a circle. In the figure below, each circle is determined by a vertex and a randomly positioned point on each adjacent side. These circles are concurrent at a point named after Auguste Miquel, who published a paper on the subject in 1838.

Tasks
1. Is it possible to position the vertices so that the Miquel point is on a vertex of the triangle? If so, under what circumstances? If not, why not?
2. Is it possible to position the vertices so that the Miquel point is on a side of the triangle? If so, under what circumstances? If not, why not?
3. Is it possible to position the vertices so that the Miquel point is outside the triangle? If so, under what circumstances? If not, why not?
4. Is it possible to position the vertices so that the Miquel point is on the circumcircle? If so, under what circumstances? If not, why not?
5. State a conjecture concerning the possible locations of the Miquel point. To prove your conjecture false, what sort of counterexample would be necessary?

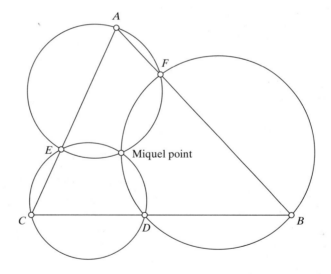

Triangle Trisection Points
Tool(s) Geometers Sketchpad   *Data File(s) cd2_2_11.gsp*

Focus
Line segments connect each vertex of a triangle to the opposite side's trisection points. This produces a variety of polygonal regions identified as follows:

1. Vertex triangles, e.g. B-Z-11
2. Vertex quadrilaterals, e.g. A-7-8-9
3. Interior triangles, e.g. 1-2-12
4. Edge pentagons, e.g. X-W-5-6-7
5. Central hexagon 6-4-2-12-10-8

Tasks
1. Write at least two conjectures about angles or angle sums.
2. Write at least two conjectures about areas or ratios of areas.
3. Write at least two conjectures about segments or ratios of segments.
4. Write at least two conjectures about perimeters or ratios of perimeters.

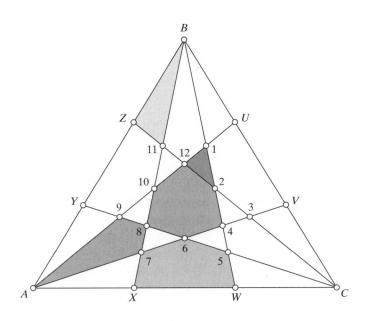

2.3   Formal Geometric Proof

- **Focus on *Informal Explanation and Proof* in the Sequence:**
  DescriptiveGeometry ► Geometric Constructions ► Observation ► Conjecture ► Testing ► Belief ► Informal Explanation ► Proof

When geometry is approached as described in sections 2.1 and 2.2 of this text, students naturally want to know if their conjectures are true. Furthermore, given the opportunity to learn why they are true, most students take a genuine interest in the logical basis of their discoveries. One way to orient students to the concept of proof is by having them read and discuss proofs related to constructions and conjectures that they have already studied. This section follows that approach by extending the insights developed in sections 2.1 and 2.2 relative to the circumcenter, incenter, orthocenter, centroid, nine-point circle, divided parallelogram, and other geometric objects.

The Circumcenter

Using the *Sketchpad* model of the circumcenter (cd2_2_1), systematic observations reveal that the circumcenter may be located anywhere inside the triangle, on the side of the triangle (only at the midpoint of the hypotenuse of a right triangle), and anywhere outside the triangle. Euclid recognized these conditions and proved a famous theorem about the circumcenter in Book IV, Proposition 5 of the *Elements.* The first part of that proof is presented in Figure 2.3.1 on the following page. The cases in which the circumcenter falls outside the triangle and on a side of the triangle are not presented. Both of these cases must be proven as well. These are left as exercises.

Because Euclid's use of language is occasionally eccentric by modern standards, a first step in "unpacking" his proofs is to rewrite them using contemporary language. This approach forces the student to deal with every thought, whether a statement (left-hand column) or a justification (right-hand column). This is precisely the sort of focus required whenever mathematical proof is done. Figure 2.3.2 on page 79 contains a rewritten version of the first part of Euclid's proof of Book IV, Proposition 5. Rewrites of the cases in which the circumcenter falls outside the triangle or on a side of the triangle are left as exercises.

Congruence of Triangles

The most subtle aspect of Euclid's Proposition 5 is its use of Proposition 4, (see Figure 2.3.3, p. 80) identified in the rewritten version of Proposition 5 as the Side-Angle-Side theorem for the congruence of two triangles. Because Proposition 4 plays a pivotal role in the proof of so many propositions, understanding it is a critical factor in understanding much of the *Elements.* Rewriting it in contemporary language is left as an exercise.

Another important means for proving two triangles congruent is presented in Euclid's Proposition 26 (see Figure 2.3.4, p. 81). In most contemporary high school geometry courses, this proposition is presented as two

**URL
2.3.1**

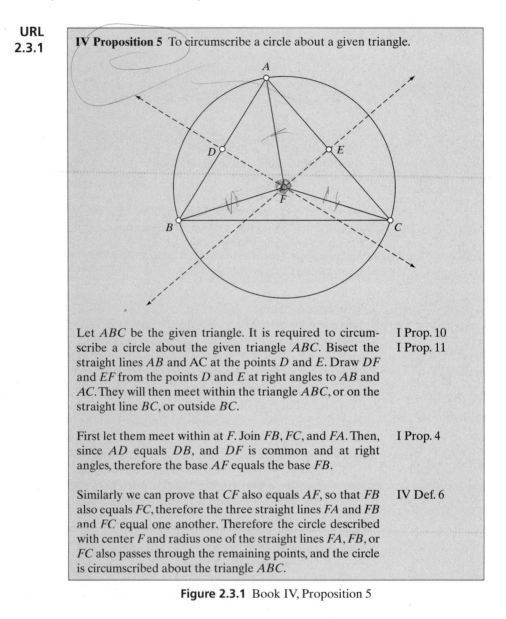

IV Proposition 5 To circumscribe a circle about a given triangle.

Let *ABC* be the given triangle. It is required to circum-   I Prop. 10
scribe a circle about the given triangle *ABC*. Bisect the   I Prop. 11
straight lines *AB* and AC at the points *D* and *E*. Draw *DF*
and *EF* from the points *D* and *E* at right angles to *AB* and
AC. They will then meet within the triangle *ABC*, or on the
straight line *BC*, or outside *BC*.

First let them meet within at *F*. Join *FB*, *FC*, and *FA*. Then,   I Prop. 4
since *AD* equals *DB*, and *DF* is common and at right
angles, therefore the base *AF* equals the base *FB*.

Similarly we can prove that *CF* also equals *AF*, so that *FB*   IV Def. 6
also equals *FC*, therefore the three straight lines *FA* and *FB*
and *FC* equal one another. Therefore the circle described
with center *F* and radius one of the straight lines *FA*, *FB*, or
FC also passes through the remaining points, and the circle
is circumscribed about the triangle *ABC*.

Figure 2.3.1 Book IV, Proposition 5

separate theorems, Angle-Angle-Side and Angle-Side-Angle. Rewriting
Proposition 26 in contemporary language is left as an exercise.

From Conjecture to Proof

As seen in Example 2.2.6, dynamic geometric software may be used to
facilitate the recognition of geometric relationships. To some extent, this
aspect of systematic exploration may be reduced to a simple process:
Measure everything you can, form ratios of measurements, observe

IV Proposition 5 How to circumscribe a circle about a given triangle.

Original Version

Let *ABC* be the given triangle. Bisect the straight lines *AB* and *AC* at the points *D* and *E*. Draw *DF* and *EF* from the points *D* and *E* at right angles to *AB* and *AC*. They will then meet within the triangle *ABC*, or on the straight line *BC*, or outside *BC*.

First let them meet within at *F*. Join *FB*, *FC*, and *FA*. Then, since *AD* equals *DB*, and *DF* is common and at right angles, therefore the base *AF* equals the base *FB*.

Similarly we can prove that *CF* also equals *AF*, so that *FB* also equals *FC*, therefore the three straight lines *FA* and *FB* and *FC* equal one another. Therefore the circle described with center *F* and radius one of the straight lines *FA*, *FB*, or *FC* also passes through the remaining points, and the circle is circumscribed about the triangle *ABC*.

Rewritten Version

Given triangle *ABC*, construct the perpendicular bisectors of sides *AB* and *AC* at points *D* and *E*, respectively. These lines will intersect at a point *F*. *F* may be located inside the triangle, outside, or on side *BC*.

Right triangles *ADF* and *BDF* are congruent by Side-Angle-Side. Consequently, $FB = FA$. Using a similar argument, it may be shown that $FC = FB = FC$, so *F* is the same distance from each vertex.

A circle circumscribes a triangle if it passes through each vertex of the triangle.

Figure 2.3.2 Rewritten Book IV, Proposition 5

which measurements or ratios of measurements remain invariant for all observed variations of the object under study, then state conjectures that generalize the invariant features. This process is teachable and readily attained by most students.

Seeing beyond these numbers to the underlying relationships that produced them requires genuine insight. Unfortunately, insight is elusive and cannot be conjured, scheduled, or taught. That being the case, one might wonder how anyone learns to prove anything. Since one aspect of insight is that of recognizing the relevance and significance of specific geometric relationships, there is a clear advantage to being in possession of as much information as possible. Generally speaking, students with a superior grasp of both the specific features of a given problem and a general knowledge of related problems and theorems are more likely to recognize useful relationships than students lacking this perspective. One way to foster this sort of knowledge is by reading and discussing Euclid's proofs.

**URL
2.3.2**

I Proposition 4 If two triangles have two sides equal to two sides respectively, and have the angles contained by the equal straight lines equal, then they also have the base equal to the base, the triangle equals the triangle, and the remaining angles equal the remaining angles respectively, namely those opposite the equal sides.

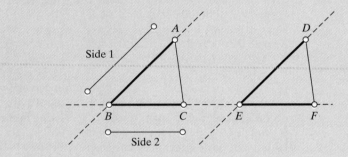

Let *ABC* and *DEF* be two triangles having the two sides *AB* and *BC* equal to the two sides *DE* and *EF* respectively, namely *AB* equal to *DE* and *BC* equal to *EF*, and the angle *ABC* equal to the angle *DEF*. I say that the base *AC* also equals the base *DF*, the triangle *ABC* equals the triangle *DEF*, and the remaining angles equal the remaining angles respectively, namely those opposite the equal sides, that is, the angle *BAC* equals the angle *EDF*, and the angle *ACB* equals the angle *DFE*. If the triangle *ABC* is superposed on the triangle *DEF*, and if the point *A* is placed on the point *D* and the straight line *AB* on *DE*, then the point *B* also coincides with the point *E* because *AB* equals *DE*.

Again, *AB* coinciding with *DE*, the straight line *AC* also coincides with *DF*, because the angle *BAC* equals the angle *EDF*. Hence the point *C* also coincides with the point *F*, because *AC* again equals *DF*. But *B* also coincides with *E*, hence the base *BC* coincides with the base *EF* and equals it.   C.N. 4

Thus the whole triangle *ABC* coincides with the whole triangle *DEF* and equals it. And the remaining angles also coincide with the remaining angles and equal them, the angle *BAC* equals the angle *EDF*, and the angle *ACB* equals the angle *DFE*.

Therefore if two triangles have two sides equal to two sides respectively, and have the angles contained by the equal straight lines equal, then they also have the base equal to the base, the triangle equals the triangle, and the remaining angles equal the remaining angles respectively, namely those opposite the equal sides.

Figure 2.3.3 Book I, Proposition 4

**URL
2.3.3**

I Proposition 26 If two triangles have two angles equal to two angles respectively, and one side equal to one side, namely, either the side adjoining the equal angles, or that opposite one of the equal angles, then the remaining sides equal the remaining sides and the remaining angle equals the remaining angle.

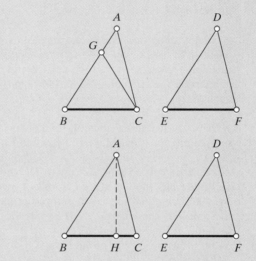

Let *ABC* and *DEF* be two triangles having the two angles *ABC* and *BCA* equal to the two angles *DEF* and *EFD* respectively, namely the angle *ABC* to the angle *DEF*, and the angle *BCA* to the angle *EFD*, and let them also have one side equal to one side, first that adjoining the equal angles, namely *BC* equal to *EF*.

I say that the remaining sides equal the remaining sides respectively, namely *AB* equals *DE* and *AC* equals *DF*, and the remaining angle equals the remaining angle, namely the angle *BAC* equals the angle *EDF*.

I Prop. 3
Axiom 1

If *AB* does not equal *DE*, then one of them is greater. Let *AB* be greater. Make *BG* equal to *DE*, and join *GC*.

Since *BG* equals *DE*, and *BC* equals *EF*, the two sides *GB* and *BC* equal the two sides *DE* and *EF* respectively, and the angle *GBC* equals the angle *DEF*, therefore the base *GC* equals the base *DF*, the triangle *GBC* equals the triangle *DEF*, and the remaining angles equal the remaining angles, namely those opposite the equal sides. Therefore the angle *GCB* equals the angle *DFE*. But the angle *DFE* equals the angle *ACB* by hypothesis. Therefore the angle *BCG* equals the angle *BCA*, the less equals the greater, which is impossible.

I Prop. 4
C.N. 1

Figure 2.3.4 Book I, Proposition 26

Therefore *AB* is not unequal to *DE*, and therefore equals it. But *BC* also equals *EF*. Therefore the two sides *AB* and *BC* equal the two sides *DE* and *EF* respectively, and the angle *ABC* equals the angle *DEF*. Therefore the base *AC* equals the base *DF*, and the remaining angle *BAC* equals the remaining angle *EDF*.

I Prop. 4

Next, let sides opposite equal angles be equal, as *AB* equals *DE*. I say again that the remaining sides equal the remaining sides, namely *AC* equals *DF* and *BC* equals *EF*, and further the remaining angle *BAC* equals the remaining angle *EDF*.

If *BC* is unequal to *EF*, then one of them is greater. Let *BC* be greater, if possible. Make *BH* equal to *EF*, and join *AH*.

I Prop. 3
Axiom 1

Since *BH* equals *EF*, and *AB* equals *DE*, the two sides *AB* and *BH* equal the two sides *DE* and *EF* respectively, and they contain equal angles, therefore the base *AH* equals the base *DF*, the triangle *ABH* equals the triangle *DEF*, and the remaining angles equal the remaining angles, namely those opposite the equal sides. Therefore the angle *BHA* equals the angle *EFD*.

I Prop. 4

But the angle *EFD* equals the angle *BCA*, therefore, in the triangle *AHC*, the exterior angle *BHA* equals the interior and opposite angle *BCA*, which is impossible.

C.N.1
I Prop. 16

Therefore *BC* is not unequal to *EF*, and therefore equals it. But *AB* also equals *DE*. Therefore the two sides *AB* and *BC* equal the two sides *DE* and *EF* respectively, and they contain equal angles. Therefore the base *AC* equals the base *DF*, the triangle *ABC* equals the triangle *DEF*, and the remaining angle *BAC* equals the remaining angle *EDF*.

Therefore if two triangles have two angles equal to two angles respectively, and one side equal to one side, namely, either the side adjoining the equal angles, or that opposite one of the equal angles, then the remaining sides equal the remaining sides and the remaining angle equals the remaining angle.

Figure 2.3.4 Book I, Proposition 26 (*continued*)

Summary

Learning to read Euclid's proofs with both understanding and appreciation cannot be accomplished in a day or a week. It is a work worthy of persistent effort over a lengthy period of time. If pursued conscientiously, the rewards are many, including...

- A deepening awareness of both the details and the overall, logical organization of the *Elements*.
- An awareness of the different approaches to proof used in the *Elements*, including proof by contradiction.
- A growing strategic sense of how to compose proofs of one's own.

Through this process, students immerse themselves in the information from which insight is constructed and create an opportunity for insight to occur. Another approach is to construct alternative proofs of propositions already studied or proofs of propositions not yet read. In any case, the *Elements* provides a rich, orderly environment in which to practice proof-making.

URLs		Note: Begin each URL with the prefix http://
2.3.1	h	http://aleph0.clarku.edu/~djoyce/java/elements/bookIV/propIV4.html
2.3.2	h	http://aleph0.clarku.edu/~djoyce/java/elements/bookI/propI4.html
2.3.3	h	http://aleph0.clarku.edu/~djoyce/java/elements/bookI/propI26.html

Table 2.3.1 Section 2.3 URLs (c = content, h = history, s = software, d = data)

Exercises

1. Rewrite the proof of each of the following propositions using contemporary language.
   a) Proposition 16   In any triangle, if one of the sides is produced, then the exterior angle is greater than either of the interior and opposite angles.
   b) Proposition 5   In isosceles triangles the angles at the base equal one another, and, if the equal straight lines are produced further, then the angles under the base equal one another.
   c) Proposition 6   If in a triangle two angles equal one another, then the sides opposite the equal angles also equal one another.
   d) Proposition 29   A straight line falling on parallel straight lines makes the alternate angles equal to one another, the exterior angle equal to the interior and opposite angle, and the sum of the interior angles on the same side equal to two right angles.

e) Proposition 30   Straight lines parallel to the same straight line are also parallel to one another.

f) Proposition 41   If a parallelogram has the same base with a triangle and is in the same parallels, then the parallelogram is double the triangle.

g) Proposition 47   In right-angled triangles the square on the side opposite the right angle equals the sum of the squares on the sides containing the right angle.

2. Present two alternative proofs of the Pythagorean Theorem. Fully explain the logic used in each case. Which of these two proofs do you prefer, and why?

3. Prove three of the conjectures that you developed in the investigations in section 2.2.

4. Prove that in any triangle, the median to a side bisects the midsegment joining the other two sides.

5. Express the radius of the incircle of a right triangle in terms of its sides, a, b, and c, where c is the hypotenuse of the triangle.

6. Determine whether two angle bisectors of a triangle be perpendicular? Two altitudes? Two medians? Justify your answers.

7. Prove that the angle between the segments from the incenter to two vertices of a triangle equals 90° plus 1/2 the measure of the angle at the third vertex.

References and Suggested Readings

Abraham, R.H. *The Visual Elements of Euclid.* Available On-Line at http://thales.vismath.org/euclid/.

Boardman, Griffin, and Murray eds. 1986. The Oxford History of the Classical World. Oxford: Oxford UP.

Cabri Geometry. Available on-line at http://www-cabri.imag.fr/index-e.html

Dunham, W. 1991. Journey Through Genius. New York: Penguin.

Euclid. Biography in Dictionary of Scientific Biography (New York 1970–1990).

Euclid. Biography in Encyclopaedia Britannica. (WWW version)

Eves, H. 1990. *An introduction to the history of mathematics.* 6th edition. Saunders College Publishing.

Fowler, D.H. 1983. *Investigating Euclid's Elements,* British J. Philos. Sci. 34, 57–70.

Fowler, D.H. 1987. *The mathematics of Plato's academy: a new reconstruction.* Oxford, 1987.

Fowler, D.H. 1990. *The mathematics of Plato's Academy: A new reconstruction.* New York, 1990.

Fraleigh, J. (1999). *A First Course in Abstract Algebra,* 6th edition. Reading, MA: Addison-Wesley.

Fraser, P.M. 1972. *Ptolemaic Alexandria* (3 vols.). Oxford, 1972

Geometers Sketchpad. Key Curriculum Press. Available on-line at www.keypress.com/sketchpad/index.html

Grattan-Guinness, I. 1996. *Numbers, magnitudes, ratios, and proportions in Euclid's Elements: How did he handle them?* Historia Math. 23 (4), 355–375.

Gray, J. 1999. *Sale of the Century?,* The Mathematical Intelligencer 21 (3), 12–15.

Hogendijk, J.P. 1987. Observations on the icosahedron in Euclid's *Elements*. Historia Math. 14 (2), 175–177.

Joyce, D. Euclid's *Elements*. Available on-line at http://aleph0.clarku.edu/~djoyce/java/elements/toc.html

Knorr, W.R. 1991. On the principle of linear perspective in Euclid's *Optics*. Centaurus 34 (3), 193–210.

Knorr, W.R. 1991. *What Euclid meant: On the use of evidence in studying ancient mathematics, in Science and philosophy in classical Greece*. New York, 1991, 119–163.

Knorr, W.R. 1992. *When circles don't look like circles: An optical theorem in Euclid and Pappus*. Arch. Hist. Exact Sci. 44 (4), 287–329.

Kreith, K. 1989. Euclid turns to probability, Internat. J. Math. Ed. Sci. Tech. 20 (3), 345–351.

Lamb, R. *Euclid and the History of Mathematics*. University of Waterloo. Available on-line at http://www.lib.uwaterloo.ca/discipline/SpecColl/euclid.html

Lindberg, D.C. 1992. *The Beginnings of Western Science*. Chicago: Chicago UP.

Martin, G. (1998). *Geometric Constructions*. New York: Springer-Verlag.

Morrow, G.W. 1992 ed. *A commentary on the first book of Euclid's Elements*. Princeton, NJ, 1992.

Mueller, I. 1981. *Philosophy of mathematics and deductive structure in Euclid's Elements*. Cambridge, Mass.-London, 1981.

National Council of Teachers of Mathematics 2000. *Principles and Standards for School Mathematics*. Available on-line at http://standards-e.nctm.org/protoFINAL/cover.html

O'Connor, J.J. & Robertson, E.F. 1999. Feuerbach. The MacTutor History of Mathematics archive. School of Mathematics and Statistics. University of St. Andrews, Scotland. Available on-line at http://www-history.mcs.st-and.ac.uk/history/Mathematicians/Feuerbach.html

Talbert, R. A. 1985. *Atlas of Classical History*. New York: Macmillan.

Theisen, W. 1984. *Euclid, relativity, and sailing*. Historia Math. 11 (1), 81–85.

Tobin, R. 1990. *Ancient perspective and Euclid's Optics*. Journal of the Warburg Courtauld Institute, 53, 14–41.

Chapter 3

Other Geometries

Imagine a sphere, vast in size but finite in diameter and volume. From your location outside the sphere, you observe that objects moving inside the sphere appear to shrink as they approach the surface of the sphere, becoming infinitesimal in the process. Conversely, objects moving toward the center of the sphere appear to grow, attaining their maximum (finite) size at the center. You also learn that, from the perspective of the inhabitants of the sphere, all meter sticks are the same length, regardless of their location.

A 2-dimensional "slice" of this universe is shown in Figure 3.1.1. Two paths are shown between points A and C, path ADC and path ABC. Which path is shorter? Since the points on path ADC are further from the center of the sphere, meter sticks laid along path ADC are shorter than meter sticks laid along path ABC. This suggests that more meter sticks might be needed along path ADC. This would mean that path ABC is shorter. By an appropriate choice of metric for this space, the shortest path from A to C may be shown to lie on an arc of a circle drawn perpendicular to the sphere at points P and Q. Under the same choice of metric, light approaching the surface slows until its speed is undetectable. In this self-contained universe, neither matter nor light ever reach the surface of the sphere.

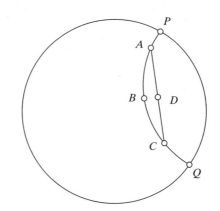

Figure 3.1.1 Non-Euclidean Shortest Paths

If such a universe were inhabited by intelligent beings, would they think it was limitless or bounded? Since light signals could never reach nor return from the surface of the sphere its presence would be undetectable by remote sensing. And because entire galaxies shrink to infinitesimal volumes near the boundary, there is no limit to the number of astronomical objects that might be contained in such a sphere. It would appear infinite in capacity from any point of view. As a result, casual observers would be expected to conclude that their universe is unbounded and infinite in size. What about our universe? Might it be bounded, like some gigantic bubble? What would you need to know to decide? While these questions are posed in a cosmological context, the questions themselves are fundamentally geometrical in nature. And the geometries are non-Euclidean. This section begins by investigating the concept of non-Euclidean spaces.

Space, Time, and Einstein

URL 3.1.1 Scientists have predicted and confirmed that rulers in our universe do change in length under certain conditions. For instance, in the Special Theory of Relativity published in 1905, Albert Einstein (1879–1955) predicted that objects traveling at velocities near the speed of light would shrink in length in the direction of motion (see Figure 3.1.2). Studies of subatomic particles have demonstrated this and other strange effects arising in situations involving relativistic speeds.

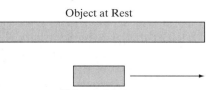

Object at Rest

Object in Motion

Figure 3.1.2 Contraction at Relativistic Speeds

URL 3.1.2 In his General Theory of Relativity published in 1915, Einstein also predicted that, under certain circumstances, light follows "shortest paths" which are not Euclidean lines. In his model of gravitation, Einstein argues that gravity is not really a force acting at a distance but an effect arising from curvatures of four dimensional spacetime, the very fabric of our universe, created by massive objects such as stars. Stated as simply as possible, "Matter tells spacetime how to curve and spacetime tells matter and light how to move." The first attempt to verify this prediction occurred in 1919 during an eclipse of the sun. Acting like a gravitational lens, the mass of the sun curved spacetime and caused light from the distant star to "bend" as it passed near the sun, changing its apparent position in the sky (see Figure 3.1.3).

Curved Spaces

Like the inhabitants of our imaginary universe, none of our perceptual senses are able to detect the bending of spacetime directly. To discover and understand such things, we must rely on proof that only logic can supply. To illustrate the sort of insights that only geometry can provide, consider the curved surface shown in Figure 3.1.4. Imagine that this surface is inhabited by an intelligent species of germ with sensory

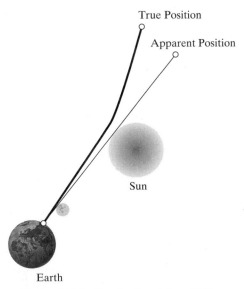

Figure 3.1.3 Gravitational Lens Effect

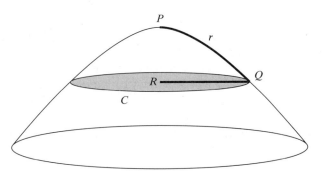

Figure 3.1.4 Circumference on a Positively Curved Surface

abilities that make them aware of their position and motion on the 2-dimensional surface but not aware of the third dimension perpendicular to the surface.

An intelligent germ located at point P may investigate the curvature of his space using some simple geometry. First, a circle with center P and radius r (segment PQ) is constructed. Next, the circumference of the circle C is measured. If the measured circumference C equals the computed circumference $\pi(2r)$, then the space is Euclidean, or "flat." As seen in Figure 3.1.4, $r > R$ and $C = \pi(2R)$ rather than $\pi(2r)$. In other words, the circumference obtained by measurement is smaller than the circumference obtained by calculation. If the germ believes his measurements, he must conclude that his space somehow curves in a direction he cannot detect directly and that his universe is not Euclidean. Spaces such as this

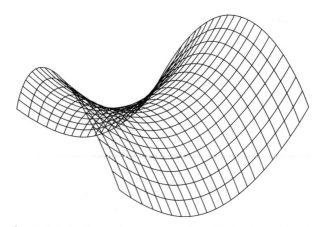

Figure 3.1.5 Circumference on a Negatively Curved Surface

are positively curved. Figure 3.1.5 shows a surface that is negatively curved. If the same procedure is carried out on this surface, the circumference obtained by measurement is larger than the circumference obtained by calculation.

While investigations of other geometries based on circumference are conceptually simple, a more formal approach offers many other interesting and powerful insights. Extending investigations to three dimensional space challenges our ability to think by analogy and our powers of visualization. An understanding of the axiomatic basis of non-Euclidean geometry can greatly facilitate these leaps of logic and belief. This chapter focuses on non-Euclidean geometries and the impact their discovery has had on the development of mathematics. We begin at the beginning, with Euclid and the concept of parallelism.

3.1   The Concept of Parallelism

In this section, you will . . .

- **Investigate the concept of logical equivalence as it applies to the parallel axiom;**
- **Review the history of the parallel axiom;**
- **Investigate the conceptual breakthrough that led to the discovery of the first non-Euclidean geometry.**

Euclid and the Concept of Parallelism

URLs
3.1.3
&
3.1.4

Axiomatic systems are formal systems of thought governed by rules of logic that operate on undefined terms, defined terms, axioms, and theorems. In creating the *Elements,* Euclid (~325–265 B.C.) created the world's first and most enduring axiomatic system. Today's logicians and mathematicians know what Euclid could only have sensed by intuition: axioms must be both consistent with and independent of one another.

Definition
3.1.1

Sets of axioms that do not contradict one another are said to be *consistent.*

Definition
3.1.2

Sets of axioms are *independent* if they may not be derived from one another.

URL
3.1.5

In comparing the content and language of Euclid's axioms, the first four are direct, concise, and easy to read. The fifth axiom seems different. Some scholars believe that this difference reflects a struggle on the part of the ancient Greeks with the concept of parallelism and its formal representation. According to Proclus (411–485 A.D.), a principal source of information about Euclid, efforts began almost immediately to show that the fifth axiom is logically dependent on the first four, that is, to prove it. One of the likely reasons for this attention is that the fifth axiom was not used in proving any of the first 28 Propositions in Book I of the *Elements.* Over the centuries, a series of geometers offered "proofs" of the fifth axiom. In each case, these arguments were shown to be flawed. These efforts were not without value, however. For instance, each of the following statements are now known to be logically equivalent to the fifth axiom, assuming the other axioms:

URL
3.1.6

- Through a point not on a given line, exactly one parallel may be drawn to the given line (John Playfair, 1748–1819).
- The sum of the angles of any triangle is equal to two right angles.
- There exists a pair of similar triangles which are not congruent.
- There exists a pair of lines everywhere equidistant from one another.
- If three angles of a quadrilateral are right angles, then the fourth angle is also a right angle.
- If a straight line intersects one of two parallel lines, it will intersect the other.
- Straight lines parallel to the same straight line are parallel to one another.
- Two straight lines which intersect one another cannot both be parallel to the same line.

If any of these statements is substituted for Euclid's fifth axiom, leaving the first four the same, the same geometry is obtained. Demonstrating this fact does not require producing all of Euclid's theorems from the alternative set of axioms. One need merely derive Euclid's axioms from the alternative set. Since the first four axioms remain the same, one need only prove Euclid's fifth axiom from the alternative set of axioms. If, in addition, the alternative set of axioms is proven starting with Euclid's, the two sets of axioms are said to be logically equivalent. The notation for this relationship is $A \Leftrightarrow B$, where A and B are logical statements, in this case sets of axioms. To prove logical equivalence, $A \Leftrightarrow B$, one must derive each set of axioms from the other, or $A \Rightarrow B$ and $B \Rightarrow A$.

Example 3.1.1

The argument presented in Table 3.1.1 establishes the logical equivalence of Euclid's fifth axiom and Playfair's version.

Saccheri

URL 3.1.7

Giovanni Girolamo Saccheri (1667–1733) was an Italian priest and mathematician. After completing his training for the Jesuit Order at the age of twenty-three, Saccheri entered his lifelong profession, university teaching. During his tenure as a professor of rhetoric, philosophy, and theology at a Jesuit College in Milan, Saccheri read Euclid's *Elements* and became fascinated by the indirect method of proof, or *reductio ad absurdum*. The method of indirect proof begins by assuming the opposite of that which one hopes to establish. Using other available information, a series of logical deductions is constructed leading eventually to a contradiction of some known fact. At that point, the assumption is known to be false, having led to a contradiction. Consequently, its negation must be true. Naturally, this approach is productive only in cases where the negation of the assumption is useful.

Just before his death, while serving as a professor of mathematics at the University of Pavia, Saccheri applied the indirect method of proof to Euclid's fifth axiom in his book *Euclid Freed of Every Flaw*. Like his many predecessors, his purpose was to derive the parallel postulate from Euclid's first four axioms. Saccheri began his argument by constructing a quadrilateral (see Figure 3.1.6) that now bears his name.

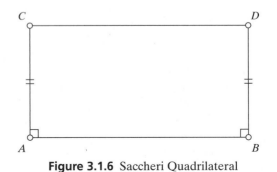

Figure 3.1.6 Saccheri Quadrilateral

Euclid ⟹ Playfair	**Playfair ⟹ Euclid**
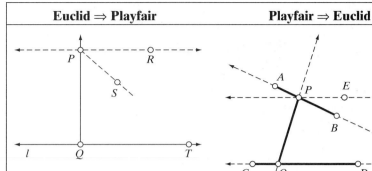	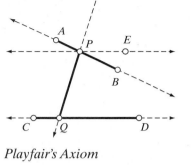

Euclid's Axiom

"If a straight line falling on two straight lines makes the interior angles on the same side less than two right angles, the two straight lines, if produced indefinitely, meet on that side on which are the angles less than the two right angles."

Let the given point be *P* and the given line be *l*. Construct a line through point *P* perpendicular to line *l* at point *Q* (Proposition 12). Construct a second line, *PR*, through *P* perpendicular to *PQ* (Proposition 11). Then lines *PR* and *l* are parallel (Proposition 27). Let *PS* be a second line through point *P* such that point *S* is in the interior of ∠*QPR*. Then ∠*QPS* < ∠*QPR* (Common Notion 5). Let *T* be a point on *l* to the right of *Q*. Then ∠*TQP* and ∠*QPS* are together less than ∠*TQP* and ∠*QPR* (Common Notion 1). Since ∠*TQP* and ∠*QPR* are both right angles, and by Euclid's axiom 5, lines *PS* and *l* intersect at some point and are therefore not parallel. So, through point *P* exactly one line may be drawn parallel to line *l*.

Playfair's Axiom

"Through a point not on a given line, exactly one parallel to the given line may be drawn."

Let lines *AB* and *CD* be cut by a line *PQ* such that ∠*DQP* and ∠*QPB* are together less than two right angles. At point *P*, construct line *PE* such that ∠*DQP* and ∠*QPE* are equal to two right angles (Proposition 23). Then *PE* is parallel to *CD* (Proposition 28). By Playfair's axiom, there is only one parallel to *CD* through point *P*, so *AB* is not parallel to *CD*. Then *AB* intersects *CD*. Assume that *AB* intersects *CD* at a point *R* on the left side of *PQ*. Then ∠*RPQ* and ∠*RQP* are together greater than two right angles. This contradicts Proposition 17, so *AB* and *CD* intersect on the right side of *PQ*. So, *AB* and *CD* intersect on the side of *PQ* on which the angles are less than two right angles.

Table 3.1.1 Logical Equivalence of Axioms

Definition 3.1.3 In Saccheri quadrilateral $ABDC$, two congruent sides CA and DB are drawn perpendicular to the base AB. Line CD is the *summit;* $\angle ACD$ and $\angle BDC$ are the *summit angles.*

Using Euclidean arguments, Saccheri proved the following theorems.

Theorem 3.1.1 The segment joining the midpoints of the summit and base of a Saccheri quadrilateral is perpendicular to both.

The proof is left to the student as an exercise.

Theorem 3.1.2 The summit angles of a Saccheri quadrilateral are congruent.

The proof is left to the student as an exercise.

Saccheri then examined three hypotheses concerning the measures of the summit angles:

Hypothesis of the acute angle The summit angles are acute.
Hypothesis of the right angle The summit angles are right angles.
Hypothesis of the obtuse angle The summit angles are obtuse.

Saccheri's plan was to show that assuming either the hypothesis of the acute angle or the hypothesis of the obtuse angle leads to a contradiction, forcing the hypothesis of the right angle to be true. This finding would imply the parallel postulate. While Saccheri succeeded in showing that the hypothesis of the obtuse angle leads to a contradiction, his effort to do the same under the hypothesis of the acute angle failed to yield a convincing contradiction. Struggling, Saccheri's effort ended in a lame argument instead of a profound discovery, the existence of an alternative to Euclidean geometry. As a consequence, he is remembered as the person who almost discovered the first non-Euclidean geometry.

Bolyai and Lobachevsky

Janos Bolyai (1802–1860) and Nicolai Lobachevsky (1792–1856) made the conceptual and psychological leap that Saccheri was unable or unwilling to accept. Bolyai, the son of a mathematics teacher, was an officer in the Hungarian army and a lifelong student of mathematics. That Bolyai realized the significance of his discovery is evident in a letter to his father in which he exclaims, "Out of nothing I have created a strange new universe." Lobachevsky was a professor of mathematics and rector of the University of Kazan, Russia. Working independently, these men approached the issue of parallelism through Playfair's form of the fifth postulate. This approach also led to three hypotheses, each of which leads to a different geometry (hyperbolic, Euclidean, and spherical, for instance).

URLs 3.1.8 & 3.1.9

Hyperbolic Axiom Through a point not on a given line, more than one parallel may be drawn to the given line.

Euclidean Axiom   Through a point not on a given line, exactly one parallel may be drawn to the given line.

Elliptic Axiom   Through a point not on a given line, no parallels may be drawn to the given line.

The geometry that they discovered was the first logically consistent alternative to Euclid. Their discovery is known as hyperbolic geometry and is characterized by the following axioms:

1. Given any two points, exactly one line may be drawn containing the points.
2. Given any line, a segment of any length may be determined on the line.
3. Given any point, a circle of any radius may be drawn.
4. All right angles are congruent.
5. Through a point not on a given line, at least two lines can be drawn that do not intersect the given line.

When first introduced to the hyperbolic axiom, most geometry students are unable to sketch an illustration. This difficulty arises because there is no obvious way to model the axiom in Euclidean space, and Euclidean space is all they know. A similar obstacle is often encountered by students of physics when challenged to "Define the term universe and give two examples."

Since the time of Bolyai and Lobachevsky, a number of models of the hyperbolic plane have been developed, including the Poincare model, the Klein model, and the Upper-Half Plane model. By embedding an infinite hyperbolic space in a finite region of the Euclidean plane, the French
URL
3.1.10
mathematician Henri Poincare (1854–1912) achieved a remarkable mathematical feat and gave students of mathematics a credible model in which to investigate the features of hyperbolic geometry.

Summary

For two thousand years, Euclid's parallel axiom was the focus of repeated efforts to show its dependence on the other four axioms of Euclidean geometry. To the mathematicians engaged in these efforts, there was only one conceivable geometry, that of Euclid. Therefore, the parallel axiom was not arbitrary, but essential. They only sought to establish the logical basis for that essential truth.

URLs
3.1.11
&
3.1.12
Bolyai and Lobachevsky looked at the same evidence as all their predecessors and reached a fundamentally different conclusion: Euclid's fifth axiom is not essential, but rather is one of three possibilities, one of which lead to hyperbolic geometry. By accepting the possibility of alternative universes, they were able to see them. Since Bolyai and Lobachevsky, geometry has enjoyed a rebirth and undergone a phenomenal revolution. Today, our view of the universe in which we live is shaped by the non-Euclidean ideas of Albert Einstein and other cosmologists. It is a good time to be a geometer!

Definitions	
3.1.1	Axioms that do not contradict one another are said to be *consistent*.
3.1.2	Axioms are *independent* if one may not be derived from another.
3.1.3	In *Saccheri quadrilateral ABDC*, two congruent *sides CA* and *DB* are drawn perpendicular to the *base AB*. Line *CD* is the *summit;* $\angle ACD$ and $\angle BDC$ are the *summit angles*.
Theorems	
3.1.1	The segment joining the midpoints of the summit and base of a Saccheri quadrilateral is perpendicular to both.
3.1.2	The summit angles of a Saccheri quadrilateral are congruent.

Table 3.1.2 Summary, Section 3.1

URLs		Note: Begin each URL with the prefix http://
3.1.1	h	www-history.mcs.st-and.ac.uk/history/Mathematicians/Einstein.html
3.1.2	h	www.rog.nmm.ac.uk/astroweb/eclipses/greenwich/1919/index.html
3.1.3	cs	aleph0.clarku.edu/~djoyce/java/elements/bookI/bookI.html#posts
3.1.4	h	www-history.mcs.st-and.ac.uk/history/Mathematicians/Euclid.html
3.1.5	h	www-history.mcs.st-and.ac.uk/history/Mathematicians/Proclus.html
3.1.6	h	www-history.mcs.st-and.ac.uk/history/Mathematicians/Euclid.html
3.1.7	h	www-history.mcs.st-and.ac.uk/history/Mathematicians/Saccheri.html
3.1.8	h	www-history.mcs.st-and.ac.uk/history/Mathematicians/Bolyai.html
3.1.9	h	www-history.mcs.st-and.ac.uk/history/Mathematicians/Lobachevsky.html
3.1.10	h	www-history.mcs.st-and.ac.uk/history/Mathematicians/Poincare.html
3.1.11	cs	www.math.uncc.edu/~droyster/courses/fall96/math3181/hypgeom.html
3.1.12	cs	www.math.ubc.ca/~robles/hyperbolic/hypr/modl/

Table 3.1.3 Section 3.1 URLs (c = concept, h = history, s = software, d = data)

Exercises

For each of the following, prove the claim that is made. You may assume Euclid's first four axioms, but not the fifth.

1. The segment joining the midpoints of the summit and base of a Saccheri quadrilateral is perpendicular to both.
2. The summit angles of a Saccheri quadrilateral are congruent.
3. In a Saccheri quadrilateral, the segment joining the midpoints of the lateral sides is perpendicular to the segment joining the midpoints of the summit and base.
4. Prove that each of the following statements is logically equivalent to Playfair's axiom:
   a) The sum of the angles of any triangle is equal to two right angles.
   b) There exists a pair of lines everywhere equidistant from one another.
   c) If three angles of a quadrilateral are right angles, then the fourth angle is also a right angle.
   d) Straight lines parallel to the same straight line are parallel to one another.

3.2   Points, Lines, and Curves in Poincare's Model of Hyperbolic Space

In this section, you will . . .

- **Use the Poincare model of the hyperbolic plane to represent points, lines, circles, and other geometric objects;**
- **Use Euclidean methods to construct orthogonal circles;**
- **Investigate the concept of parallelism in the hyperbolic plane;**
- **Compute the distance between hyperbolic points;**
- **Identify similarities between Euclidean and hyperbolic geometry.**

Hyperbolic Geometry: Points

URL 3.2.1

In Poincare's disc model of the hyperbolic plane, Euclidean points lying within the unit circle are used to represent points in the hyperbolic plane. For instance, in Figure 3.2.1 Euclidean points C and D are also points of the hyperbolic plane. While Euclidean points A and B are not points of the hyperbolic plane, they are treated as points-at-infinity, or vanishing points. In general, all such points are called omega (Ω) points. Figure 3.2.2 shows a triangle formed by joining Ω points A, B, and C.

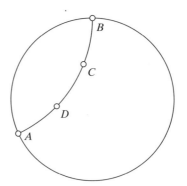

Figure 3.2.1 Distance in the Poincare Model

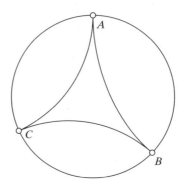

Figure 3.2.2 Omega Points

Definition 3.2.1

Given a unit circle Σ in the Euclidean plane, *points* of the hyperbolic plane are Euclidean points in the interior of Σ.

**Definition
3.2.2**
Given a unit circle Σ in the Euclidean plane, *omega* points (Ω) of the hyperbolic plane are Euclidean points on Σ.

**URL
3.2.2**
Hyperbolic Geometry: Lines
In Poincare's model, lines consist of arcs of circles drawn orthogonal to Σ.

**Definition
3.2.3**
Given a unit circle Σ in the Euclidean plane, *lines* of the hyperbolic plane are arcs of circles drawn orthogonal to Σ and located in the interior of Σ.

The following information provides a method for sketching and/or constructing curves with this feature. Figure 3.2.3 shows two orthogonal circles with centers P and Q and intersection points A and B. The steps used to construct this figure are

1. Draw circle P and ray PA as shown.
2. Construct a line perpendicular to ray PA at point A.
3. Select an arbitrary point Q on the perpendicular line as the center of the second circle.
4. Use segment QA as the radius of the second circle.
5. Construct the second circle and mark the second point of intersection as point B.

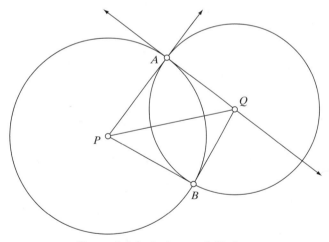

Figure 3.2.3 Orthogonal Circles

**Theorem
3.2.1**
If two circles are orthogonal at one point of intersection, they are orthogonal at the other point of intersection as well (see Figure 3.2.3).

Given that PA is perpendicular to QA, show that PB is perpendicular to QB. The details of this proof are left as an exercise.

The following methods are useful in constructing orthogonal circles. While the mathematical basis of each method is discussed briefly, an understanding of these methods is not critical in developing an understanding of the content of this chapter.

Given points *A* and *B* on Σ, construct rays *PA* and *PB* and the lines perpendicular to these rays at points *A* and *B* (see Figure 3.2.4). Let *Q* be the point of intersection of these perpendicular lines. Use *QA* as the radius of circle *Q*. Show that these circles are orthogonal.

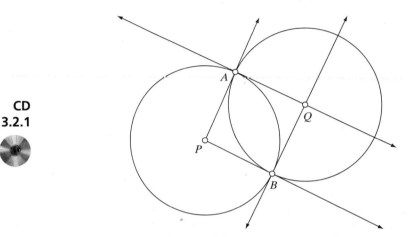

Figure 3.2.4 Two Points on the Circle

Given one point *A* on Σ and a second point *B* in the interior of Σ, draw ray *PA* and construct the line perpendicular to ray *PA* at point *A* (see Figure 3.2.5). Draw segment *AB* and construct the perpendicular bisector of segment *AB*. Let *Q* be the intersection of these two perpendicular lines. Use *QA* as the radius of circle *Q*. Show that these circles are orthogonal.

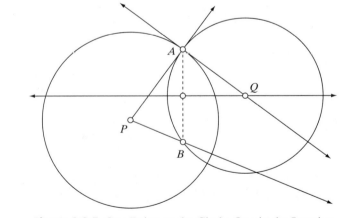

Figure 3.2.5 One Point on the Circle, One in the Interior

Given two points *A* and *B* in the interior of Σ, a third point *C* on the second circle is needed to determine its center (see Figure 3.2.6). Any one of the following three methods may be used to identify the third point, *C*.

**CD
3.2.3**

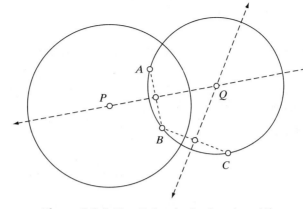

Figure 3.2.6 Two Points in the Interior of Σ

Method 1: Based on inversion of the circle (see Figure 3.2.7). Position point C so that $PC*PB = r^2$.

**CD
3.2.4**

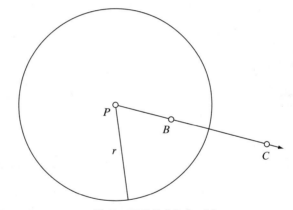

Figure 3.2.7 Method 1

Method 2: Based on a complete quadrilateral (see Figure 3.2.8). Locate point V_1 as shown in Figure 3.2.8. Draw rays LV_1, RV_1 and BV_1. Draw ray LV_4 and label its intersections with rays RV_1 as V_4 and with BV_1 as V_3. Draw ray RV_3 and label its intersection with ray LV_1 as V_2. Point C is located at the intersection of rays PB and V_4V_2.

Method 3: Based on harmonic sets (see Figure 3.2.9). Position point C so that

$$\frac{1}{LB} - \frac{1}{LR} = \frac{1}{LR} - \frac{1}{LC}$$

**URL
3.2.3**

Hyperbolic Geometry: Distance

Because the area within the Euclidean unit circle represents an infinite hyperbolic plane, most of the hyperbolic plane is enormously compressed. This means that identical Euclidean segments positioned at

CD
3.2.5

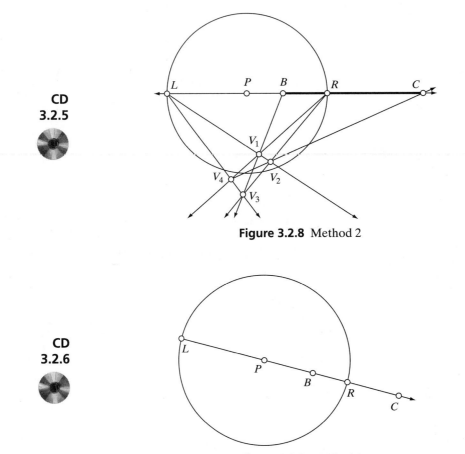

Figure 3.2.8 Method 2

CD
3.2.6

Figure 3.2.9 Method 3

different locations within the circle have different lengths in the hyperbolic plane. This concept is expressed formally in the relationship

$$d\rho = \frac{2\,dr}{1 - r^2}$$

where ρ represents ruler length in hyperbolic distance and r is the Euclidean distance from the center of the unit circle. An examination of this expression reveals that $d\rho \to \infty$ as $r \to 1$. In other words, as rulers approach the unit circle, they represent larger and larger distances. Alternatively, the amount of Euclidean space required to represent a given area of hyperbolic space shrinks as one approaches the unit circle.

The relationship between the Euclidean distance of a point from the center of the Poincare disk and its corresponding hyperbolic distance may be expressed as

$$\rho = \int_0^r \frac{2\,dt}{1 - t^2} = 2\,\tanh^{-1} r, \quad \text{or} \quad r = \tanh\frac{\rho}{2}$$

Definition 3.2.4   The Euclidean distance and hyperbolic distance of a point from the center of the Poincare disk are related by the formula

$$r = \tanh \frac{\rho}{2}$$

For measuring distances between points located on the same infinite hyperbolic line, the following formulas are preferred by many for their simplicity.

Definition 3.2.5   If a Euclidean line determined by the two hyperbolic points C and D intersects Σ in the Euclidean points A and B, the hyperbolic *distance* from C to D is given by

$$d(C, D) = \log\left[\frac{CA/CB}{DA/DB}\right]$$

where CA, CB, DA, and DB represent Euclidean arc lengths. This equation may also be written as $d(C, D) = \log|(CD, AB)|$, where (CD, AB) is the *cross ratio* of C and D with respect to A and B and

$$(CD, AB) = \left[\frac{CA/CB}{DA/DB}\right]$$

Definition 3.2.5 may be used to provide an alternative method to that presented in Definition 3.2.4.

Theorem 3.2.2   If a point A in the interior of Σ is located a Euclidean distance $r < 1$ from the center O, its hyperbolic distance from the center is given by

$$d(A, O) = \log \frac{1 + r}{1 - r}$$

A proof of this theorem is left as an exercise.

Theorem 3.2.3   The hyperbolic distance from any point in the interior of Σ to the circle itself is infinite.

A proof of this theorem is left as an exercise.

Example 3.2.1 illustrates the use of the distance formula in Definition 3.2.5.

Example 3.2.1   Table 3.2.1 is based on Figure 3.2.1. Arc lengths CA, CB, DA, and DB are Euclidean distances. CD is the computed hyperbolic distance.

Hyperbolic Geometry: Parallel Lines

Using Poincare's model, the hyperbolic axiom may be illustrated directly (see Figure 3.2.10). Given line AB and point D not on AB, lines CD and DE fail to intersect AB. Since neither of these lines intersect AB, they are both parallel to AB. It is important at this juncture to recognize that lines CD and DE are geodesics, or shortest paths, in the hyperbolic plane and correspond to the paths that light would follow in moving from point

CA	CB	DA	DB	CD	Hyperbolic CD
1.0	.8	.1	1.7	.9	$\log \dfrac{1.0/.8}{.1/1.7} = 1.327$
1.0	.8	.01	1.79	.99	$\log \dfrac{1.0/.8}{.01/1.79} = 2.350$
1.0	.8	.001	1.799	.999	$\log \dfrac{1.0/.8}{.001/1.799} = 3.352$
1.0	.8	.0001	1.7999	.9999	$\log \dfrac{1.0/.8}{.0001/1.7999} = 4.352$
1.0	.8	.00001	1.7999	.99999	$\log \dfrac{1.0/.8}{.00001/1.79999} = 5.352$

Table 3.2.1 Hyperbolic Distance

to point. To the inhabitants of such a space, these lines would look as straight as line AB.

In addition to lines such as CD and DE in Figure 3.2.10, there is a second class of lines that are considered parallel to line AB. [Note: For simplicity, the following definition is constructed with reference to the Poincare model and is not transferable to other models of hyperbolic geometry.]

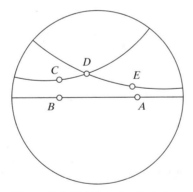

Figure 3.2.10 Hyperbolic Axiom

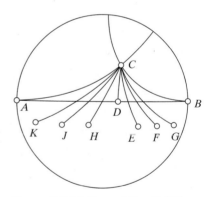

Figure 3.2.11 Sensed-Parallel Lines

Definition 3.2.6 If two Euclidean circles orthogonal to Σ intersect at a point on Σ, the hyperbolic lines determined by their points in the interior of Σ are said to be *sensed-parallel*.

For instance, in Figure 3.2.11, line CD is drawn perpendicular to line AB. A series of lines is then drawn through point C that make increasingly greater angles with line CD, measured in a counterclockwise direction: CE, CF, CG. Each of these lines intersects hyperbolic line AB in a hyperbolic point.

Definition
3.2.7

The first such line not to intersect *AB* in this manner is line *BC*, which intersects Euclidean line *AB* at point *B,* an Ω point. Hyperbolic line *BC* is said to be *right-sensed parallel* to line *AB*.

Definition
3.2.8

Applying a similar procedure on the other side of line *CD*, the first line measured in a clockwise direction from *CD* that does not intersect *AB* is line *AC*. This line is said to be *left-sensed parallel* to line *AB*.

Sensed-parallel lines such as *AC* and *BC* are a distinctive feature of hyperbolic geometry. No comparable feature exists in Euclidean geometry. Figure 3.2.12 shows line *AB*, sensed-parallel lines *AC* and *BC*, and the perpendicular to line *AB*, segment *CD*. Angles *ACD* and *BCD* are called *angles of parallelism*.

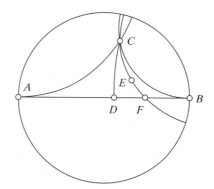

Figure 3.2.12 Angle of Parallelism: Congruence

Theorem
3.2.4

The angles of parallelism associated with a given line and point are congruent.

In Figure 3.2.12, assume that ∠*ACD* ≠ ∠*BCD*. Then one angle is greater than the other. Let ∠*ACD* < ∠*BCD*. Then there is a point *E* in the interior of ∠*BCD* such that ∠*ACD* = ∠*ECD*. Line *EC* must intersect line *AB* since *BC* is a right-sensed parallel line to *AB*. Let the intersection of *EC* and *AB* be the point *F*. If point *G* is located between points *D* and *A* such that *DG* = *DF*, then △*GCD* is congruent to △*FCD*. (Why?). As a result, ∠*HCD* = ∠*FCD*. But ∠*FCD* = ∠*ECD* = ∠*ACD*. Then *AC* intersects *AB* in point *H*. Since we know that *AC* is left-sensed parallel to *AB*, this is a contradiction. So ∠*ACD* = ∠*BCD* and the angles of parallelism are congruent.

Theorem
3.2.5

The angles of parallelism associated with a given line and point are acute.

In Figure 3.2.13, assume that ∠*DCF*, the angle of parallelism, is greater than 90°. Then we can construct ∠*DCE* = 90°. Since both *AB* and *CE* are perpendicular to *CD*, they are parallel. Then *CE* doesn't intersect *AB*. This is a contradiction that *CF* is the first line through *C* not to intersect AB. So the angle of parallelism cannot be greater than 90°. If the angle of parallelism is 90°, the left-sensed and right-sensed parallel lines coincide and the geometry becomes Euclidean rather than hyperbolic. (Why?) Having eliminated the other possibilities, the angle of parallelism must be acute.

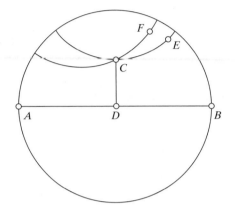

Figure 3.2.13 Angle of Parallelism: Acute

Theorem 3.2.6 Given a point P a hyperbolic distance d from a hyperbolic line AB, the angle of parallelism, τ, with respect to this line is given by (Lobachevsky's Theorem)

$$e^{-d} = \tan\left(\frac{\tau}{2}\right)$$

A line is drawn perpendicular to line AB through point P, intersecting line AB at point R (see Figure 3.2.14). This line also contains the center of Σ and Q, the center of the Euclidean arc that is the hyperbolic line AB. Using a hyperbolic transformation similar to a slide in Euclidean space, point P and line AB are moved relative to this line until point P coincides with the center of Σ. This transformation preserves the distance from the point to the line and all associated angles. Next, radii QA and QB are drawn perpendicular, respectively, to radii PA and PB. Hyperbolic line PA is right-sensed parallel to line AB, with τ the angle of parallelism. From Theorem 3.2.2, if r is the Euclidean distance from P to R, the corresponding hyperbolic distance is

$$d = \log \frac{1+r}{1-r}$$

This expression may be rewritten using exponents as

$$e^{d} = \frac{1+r}{1-r} \Rightarrow e^{-d} = \frac{1-r}{1+r}$$

Using the geometry of Euclidean right triangles and noting that $QA = QR$, we obtain

$$r = \sec \tau - \tan \tau = \frac{1 - \sin \tau}{\cos \tau}$$

Substitution leads to

$$e^{-d} = \frac{1-r}{1+r} = \frac{\cos \tau + \sin \tau - 1}{\cos \tau - \sin \tau + 1} = \frac{\cos^2 \tau + 2 \cos \tau \sin \tau + \sin^2 \tau - 1}{\cos^2 \tau + 2 \cos \tau - \sin^2 \tau + 1}$$

$$= \frac{2 \cos \tau \sin \tau}{2 \cos^2 \tau + 2 \cos \tau} = \frac{\sin \tau}{\cos \tau + 1} = \frac{2 \sin\left(\frac{\tau}{2}\right) \cos\left(\frac{\tau}{2}\right)}{\left(2 \cos^2\left(\frac{\tau}{2}\right) - 1\right) + 1} = \tan\left(\frac{\tau}{2}\right)$$

Example 3.2.2   Find the angle of parallelism τ associated with line AB in Figure 3.2.14 given that $PR = .25$.

$$e^{-.25} \Rightarrow .7788 \Rightarrow \tau/2 = \arctan(.7788) = 37.9° \Rightarrow \tau = 75.8°$$

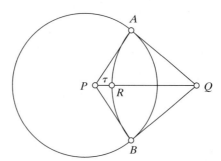

Figure 3.2.14 Angle of Parallelism and Distance

Example 3.2.3   Find the distance of a point P from a line l if the angle of parallelism through P relative to line l is 89°.

$$\tan \frac{89}{2} = .9827 \Rightarrow e^{-d} = .9827 \Rightarrow d = .017$$

Example 3.2.4   Find the distance of a point P from a line l if the angle of parallelism through P relative to line l is 2°.

$$\tan \frac{2}{2} = .017 \Rightarrow e^{-d} = .017 \Rightarrow d = 4.07$$

Hyperbolic Geometry: Circles

Considering the fact that rulers vary in size with their distance from the center of Σ, one might expect circles to be elliptical or ovoid in form. This turns out not to be the case. Instead, hyperbolic circles embedded in Euclidean space retain their circular appearance, with offset centers. For

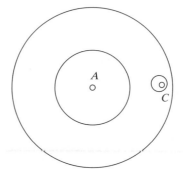

Figure 3.2.15 Hyperbolic Circles

instance, circle A and circle C in Figure 3.2.15 have identical radii, and therefore identical circumferences and areas. The center of circle A is the same as the center of Σ, so there is no apparent offset. Although the center of circle C may seem closer to the circle on the right side than on the left, in hyperbolic space, the distances are the same. Because it is an infinite distance from any point C to Σ, a circle of any radius could be drawn with C as its center. Using calculus, the circumference C of a hyperbolic circle with radius R may be determined.

Theorem 3.2.7   Given a hyperbolic circle with radius R, the circumference C of the circle is given by $C = 2\pi \sinh(R)$.

Example 3.2.5   Find the circumference of a hyperbolic circle with radius 1. $C = 2\pi \sinh(1) = 7.38$. Notice that the circumference of this hyperbolic circle is greater than the circumference of a Euclidean circle with the same radius.

Similarities Between Euclidean and Hyperbolic Geometry

URL 3.2.4   Because the first four axioms are the same in both Euclidean and Hyperbolic geometry, any similarities and differences between the two geometries arise from differences in the fifth axioms. Since Euclid did not use the fifth axiom in his first 28 Propositions in *Book I* of the *Elements*, each of these is also true in hyperbolic space. For instance, Table 3.2.2 shows several examples in which selected propositions are illustrated both in Euclidean and hyperbolic figures. Illustrating the rest of these 28 Propositions is left as an exercise.

Proposition 1 To construct an equilateral triangle on a given finite straight line.

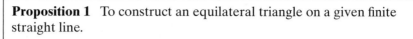

Proposition 4 If two triangles have two sides equal to two sides respectively, and have the angles contained by the equal straight lines equal, then they also have the base equal to the base, the triangle equals the triangle, and the remaining angles equal the remaining angles respectively, namely those opposite the equal sides.

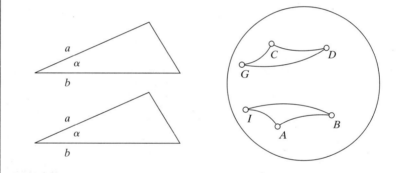

Proposition 12 To draw a straight line perpendicular to a given infinite straight line from a given point not on it.

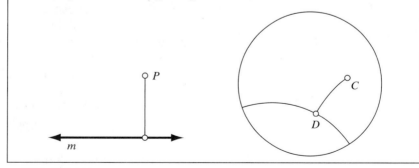

Table 3.2.2 Selected Axioms True in Euclidean and Hyperbolic Geometry

Summary

The Poincare disk model of hyperbolic space is one of three used widely by mathematicians, the other two being the Klein model and the Upper Half Plane model. The basic features of the Poincare model are summarized in Table 3.2.3.

Definitions	
3.2.1	Given a unit circle Σ in the Euclidean plane, *points* of the hyperbolic plane are Euclidean points in the interior of Σ.
3.2.2	Given a unit circle Σ in the Euclidean plane, *omega points* (Ω) of the hyperbolic plane are Euclidean points on Σ.
3.2.3	Given a unit circle Σ in the Euclidean plane, *lines* of the hyperbolic plane are arcs of circles drawn orthogonal to Σ.
3.2.4	The Euclidean distance and hyperbolic distance of a point from the center of the Poincare disk are related by the formula $$r = \tanh \frac{\rho}{2}$$
3.2.5	If a Euclidean line determined by the two hyperbolic points C and D intersects in the Euclidean points A and B, the hyperbolic *distance* from C to D is given by $$d(C, D) = \log\left[\frac{CA/CB}{DA/DB}\right]$$
3.2.6	If two Euclidean circles orthogonal to Σ intersect at a point on Σ, the hyperbolic lines determined by their points in the interior of Σ are said to be *sensed-parallel*.
3.2.7	Line CD is drawn perpendicular to line AB. A series of lines is then drawn through point C that make increasingly greater angles with line CD, measured in a counterclockwise direction: CE, CF, CG. Each of these lines intersects hyperbolic line AB in a hyperbolic point. The first such line not to intersect AB in this manner is line BC, which intersects Euclidean line AB at point B, an Ω point. Hyperbolic line BC is said to be *right-sensed parallel* to line AB.
3.2.8	Applying a similar procedure on the other side of line CD, the first line measured in a clockwise direction from CD that does not intersect AB is line AC. This line is said to be *left-sensed parallel* to line AB.

Table 3.2.3 Summary, Section 3.2

Theorems

3.2.1	If two circles are orthogonal at one point of intersection, they are orthogonal at the other point of intersection as well.
3.2.2	If a point A in the interior of Σ is located a Euclidean distance $r < 1$ from the center O, its hyperbolic distance from the center is given by

$$d(A, O) = \log \frac{1 + r}{1 - r}$$

3.2.3	The hyperbolic distance from any point in the interior of Σ to the circle itself is infinite.
3.2.4	The angles of parallelism associated with a given line and point are congruent.
3.2.5	The angles of parallelism associated with a given line and point are acute.
3.2.6	Given a point P a hyperbolic distance d from a hyperbolic line AB, the angle of parallelism, τ, with respect to this line is given by (Lobachevsky's Theorem)

$$e^{-d} = \tan\left(\frac{\tau}{2}\right)$$

3.2.7	Given a hyperbolic circle with radius R, the circumference C of the circle is given by $C = 2\pi \sinh(R)$.

Table 3.2.3 Summary, Section 3.2 (*continued*)

URLs		Note: Begin each URL with the prefix http://
3.2.1	cs	math.rice.edu/~joel/NonEuclid/
3.2.2	s	www.math.ubc.ca/~robles/hyperbolic/hypr/modl/pncr/pncrjava.html
3.2.3	c	www.math.ubc.ca/~robles/hyperbolic/hypr/modl/pncr/eq.html
3.2.4	cs	aleph0.clarku.edu/~djoyce/java/elements/toc.html

Table 3.2.4 Section 3.2 URLs (c = concept, h = history, s = software, d = data)

Exercises

Sketch

1. Sketch hyperbolic line *AB* given the following information:

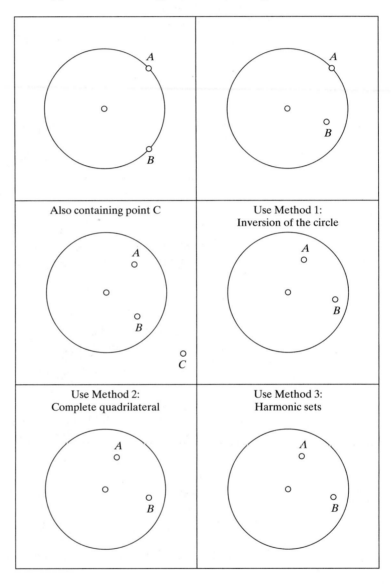

2. Triangles with the following features:

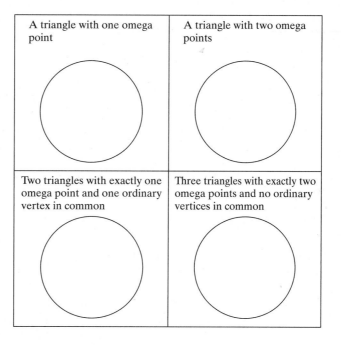

A triangle with one omega point	A triangle with two omega points
Two triangles with exactly one omega point and one ordinary vertex in common	Three triangles with exactly two omega points and no ordinary vertices in common

3. Both Euclidean and hyperbolic illustrations of the first 28 Propositions of Euclid's Elements not addressed in Table 3.2.2.

Find

4. The hyperbolic distance from C to D (shown below) given the following Euclidean distances: $CA = 1.4, CB = .4, DA = .3, DB = 1.5$

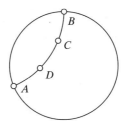

5. Find the circumference of hyperbolic circles with radius 2. With radius 4.

Prove the following

6. If two circles are orthogonal at one point of intersection, they are orthogonal at the other point of intersection as well.

7. If a point A in the interior of S is located a Euclidean distance $r < 1$ from the center O, its hyperbolic distance from the center is given by

$$d(A, O) = \log \frac{1 + r}{1 - r}$$

8. The hyperbolic distance from any point in the interior of Σ to the circle itself is infinite.

Investigations

Orthogonal Circles
First Case: One Point on the Circle and One Point in the Interior of the Circle
Tool(s) Geometers Sketchpad   *Data File(s)* cd3_2_2.gsp

Focus
Circle *P* is constructed orthogonal to Circle *O* given one point on Circle *O* and a given point in the interior of Circle *O*.

Tasks
1. Construct a second set of tangents to both circles at their other point of intersection. What angles are formed? When two circles are found to be orthogonal at one point of intersection, will they always be orthogonal at the other point of intersection? Why or why not?
2. Describe what happens when you move the interior drag pt. outside of Circle *O*. Is this construction valid whether the given point is in the interior or exterior of the circle? Why or why not?

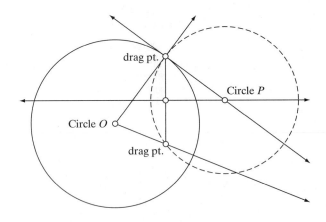

Second Case: Two Points in the Interior of the Circle
Tool(s) Geometers Sketchpad   *Data File(s)* cd3_2_3.gsp

Focus
Circle *P* is constructed orthogonal to Circle *O* given two points in the interior of Circle *O*.

Tasks
1. What happens to Circle *P* when the given points move closer to Circle *O*? To the center of Circle *O*?

2. Describe what happens when you move the given points outside of Circle *O*. Does the construction appear to be valid whether the given points are in the interior or exterior of the Circle *O*?

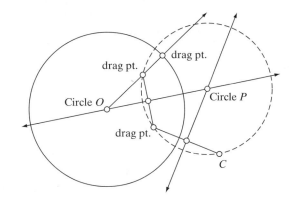

Orthogonal Circles from Harmonic Sets
Tool(s) Geometers Sketchpad   *Data File(s)* cd3_2_5.gsp

Focus
Point *D* is given as one of the points in the interior of Circle *O* through which orthogonal Circle *P* must pass. *AB*, a diameter of Circle *O*, is drawn containing point *D*. The task is to find a point *C* on line *AB* that is also on Circle *P*.

- Four points, *A, B, C*, and *D* form a harmonic set *H*(*AB, CD*) if there is a complete quadrangle in which two opposite sides pass through *A*, two other opposite sides pass through *B*, and the remaining two sides pass through *C* and *D*, respectively. In gsp16, the six lines defined by quadrangle *PQRS* intersect the diameter of Circle *O* at points *A, B, C*, and *D*, guaranteeing that these points form a harmonic set. Adjusting the location of point *D* on this diameter results in a new position for point *C* as determined by quadrangle *PQRS*.
- If two intersecting circles divide a line segment containing the center of one of the circles harmonically (cross ratio = −1), then the circles are orthogonal.

Tasks
1. Verify that the circles are orthogonal by constructing tangents at the points of intersection.
2. Move point *D* along line *AB*. State a conjecture concerning the movement of point *C*. To prove your conjecture false, what sort of counter example would be necessary?
3. Using the arrow icon, resize Circle *O* using the drag pt. Discuss your observations.
4. Using the arrow icon, slide the center of Circle *P* along the secant containing its diameter. Discuss your observations.

5. Why is the movement of the center of Circle *P* constrained? Discuss why this procedure generates many orthogonal circles instead of just one.
6. Move point *D* outside of Circle *O*. Discuss your observations.
7. State a conjecture concerning the possible locations of the center of Circle *P* (shown below). To prove your conjecture false, what sort of counter example would be necessary?

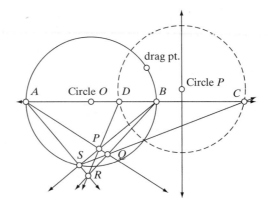

Introduction to Non-Euclid
Tool(s) Non-Euclid   *Data File(s) none*

Focus
Non-Euclid includes a set of introductory activities that highlight the basic features of hyperbolic geometry as demonstrated on the Poincare disk.

Tasks
The following activities are available under the Help pull-down menu. Work through each activity looking for answers to the following questions.

1. In any geometry, the term "line" is reserved for the shortest curve between two points. What is the apparent angular relationship between lines in the Poincare disk and the boundary of the disk?
2. If you lived in a hyperbolic universe, would you believe it to be finite or infinite? Design an experiment that would tell you whether your space was "flat" or "curved."
3. If "parallel" lines are defined as nonintersecting, how many parallels may be drawn to a given line through a point not on the given line?
4. Using Non-Euclid, investigate possible angle sums for hyperbolic triangles. What is the minimum possible angle sum? Under what conditions do such triangles occur? Is there a maximum possible angle sum?
5. Construct several right triangles and measure their sides. Using a calculator, determine whether the Pythagorean Theorem holds in hyperbolic geometry.

6. Is it possible to construct a rectangle in hyperbolic geometry? Why or why not?

7. How is area measured in hyperbolic geometry? Why are conventional units of area impossible in hyperbolic space?

8. Construct several unit circles within the Poincare disk. How does the appearance of circles drawn near the boundary of the disk differ from circles drawn near the center of the disk? If you could move a ruler from the center of the disk toward the boundary, what changes would you observe in the ruler?

9. How does hyperbolic geometry differ from Euclidean geometry? Using Non-Euclid, construct hyperbolic counter examples to the following theorems in Euclidean geometry.

   - If two lines are parallel to a third line, they are parallel to each other.
   - If two parallel lines are cut by a transversal, the alternate interior angles are congruent and the corresponding angles are congruent.
   - The measure of an exterior angle of a triangle is equal to the sum of the measures of the remote interior angles.
   - Any angle inscribed in a semicircle is a right angle.
   - The altitude to the hypotenuse of a right triangle forms two triangles, each of which is similar to the original triangle and to each other.

10. How is hyperbolic geometry similar to Euclidean geometry? Using Non-Euclid, illustrate the following theorems common to both geometries. Sketch the results in the space provided.

    - You can construct an equilateral triangle.
    - The base angles of an isosceles triangle are congruent.
    - Vertical angles are congruent.
    - Angle-Side-Angle establishes congruence between triangles.
    - Angle-Angle-Side establishes congruence between triangles.

3.3 Polygons in Hyperbolic Space

Saccheri Quadrilaterals

Theorems 3.1.1 and 3.1.2 established that "The segment joining the midpoints of the summit and base of a Saccheri quadrilateral is perpendicular to both" and "The summit angles of a Saccheri quadrilateral are congruent." These findings are true in Euclidean, hyperbolic, and elliptic geometry. In each of these geometries, additional features reflect the consequences of the Euclidean, hyperbolic, or elliptic axiom, respectively. For instance, Figure 3.3.1 shows a Saccheri quadrilateral $ABEF$ using the Poincare model of hyperbolic geometry. Appearances seem to indicate that, in hyperbolic space, the summit angles of a Saccheri quadrilateral are acute. These conjectures are formalized in Theorem 3.3.1.

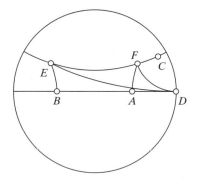

Figure 3.3.1 Summit Angles Acute

Theorem 3.3.1

The summit angles of a Saccheri quadrilateral are acute.

Figure 3.3.1 shows a Saccheri quadrilateral in the Poincare model of hyperbolic space. Point D is an Omega point associated with hyperbolic line AB. Lines ED and FD are right-sensed parallel to AB from points E and F, respectively. Since $EB = FA$, $\angle BED = \angle AFD$. Why? By Theorem 3.1.2, $\angle BED + \angle DEC = \angle AFE$. Because they form a straight angle, $\angle AFE + \angle AFD + \angle DFC = 180°$. By substitution, $(\angle BED + \angle DEC) + (\angle BED) + \angle DFC = 180°$. By the exterior angle theorem and relative to $\triangle DEF$, $\angle DEC < \angle DFC$. Then $2\angle BED + 2\angle DEC < 180° \Rightarrow \angle BED + \angle DEC < 90°$. So, the summit angles of the Saccheri quadrilateral are acute.

Triangle Angle Sums

Using the fact that the summit angles of a Saccheri quadrilateral are acute, we now proceed to prove one of the most famous features of hyperbolic space.

Theorem 3.3.2

The angle sum of hyperbolic triangles is less than 180°.

Three cases need to be considered (see Figure 3.3.2).

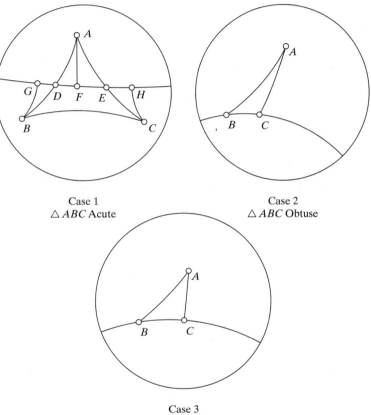

Case 1
△ *ABC* Acute

Case 2
△ *ABC* Obtuse

Case 3
△ *ABC* Right

Figure 3.3.2 Triangle Angle Sum

Case 1: △*ABC* is acute.
Points *D* and *E* are midpoints of segments *AB* and *AC*, respectively. Segments *AF*, *BG*, and *CH* are drawn perpendicular to line *DE*. ∠*BDG* = ∠*ADF*. Then △*BGD* ≅ △*ADF* by angle-angle-side. Using a similar argument, △*CEH* ≅ △*AEF*. Then ∠*GBD* = ∠*FAD* and ∠*FAE* = ∠*HCE*. By substitution, ∠*ABC* + ∠*BCA* + ∠*CAB* = ∠*GBC* + ∠*HCB*. By Common Notion 1, *BG* = *CH*. Then *GHCB* is a Saccheri quadrilateral. Consequently, the angle sum of △*ABC* is the same as the sum of the summit angles of

Saccheri quadrilateral $GHCB$. Since the summit angles of a Saccheri quadrilateral are acute, the angle sum of $\triangle ABC$ is less than $180°$.

Case 2: $\triangle ABC$ is obtuse
The proof is left as an exercise.

Case 3: $\triangle ABC$ is right.
The proof is left as an exercise.

Since triangles have angle sums less than $180°$, the amount by which the angle sum of a given triangle falls short of $180°$ has been given a name.

Definition 3.3.1

The *defect* of a hyperbolic triangle with degree angle sum σ is $180° - \sigma$. The *defect* of a hyperbolic triangle with radian angle sum σ is $\pi - \sigma$.

Theorem 3.3.3

The sum of the acute angles of a right hyperbolic triangle is less than $90°$.

This result follows directly from Theorem 3.3.2.

Since hyperbolic triangles have angle sums less than $180°$, it follows that the Euclidean formula for the sum of the interior angles of a convex n-gon is invalid in hyperbolic space.

Theorem 3.3.4

The sum of the interior angles of a convex hyperbolic polygon is less than $(n - 2)180°$.

The proof of this theorem is left as an exercise.

Theorem 3.3.5

The angle sum of a Saccheri quadrilateral is less than $360°$.

A proof of this theorem is left as an exercise.

Theorem 3.3.6

Rectangles do not exist in hyperbolic space.

A proof of this theorem is left as an exercise.

The Hyperbolic Pythagorean Theorem

Figure 3.3.3 shows a hyperbolic right triangle with $AC \perp BC$, $AB = 1.81$, $AC = 1.24$, and $BC = 1.11$. Since $1.81^2 \neq 1.24^2 + 1.11^2$, the Euclidean Pythagorean Theorem is not valid in hyperbolic space. Using techniques beyond the scope of this text, the following hyperbolic Pythagorean Theorem may be proven.

Theorem 3.3.7

In hyperbolic right triangle ABC, sides a, b, and c are opposite angles A, B, and C, respectively. Angle C is the right angle. Then $\cosh(c) = \cosh(a)\cosh(b)$.

Example 3.3.1

In Figure 3.3.3, let $AC \perp BC$, $AB = 1.81$, $AC = 1.24$, and $BC = 1.11$. Ignoring round-off inaccuracies, $\cosh(1.81) = 3.14$ and $\cosh(1.24)\cosh(1.11) = 3.14$.

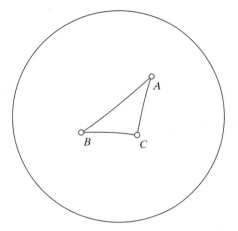

Figure 3.3.3 Hyperbolic Pythagorean Theorem

Distance Between Lines

Euclid's Proposition 12 states that a line may be drawn from a point not on a given line perpendicular to a given line. This is a theorem of hyperbolic space as well as Euclidean space (why?). Since the distance from a point to a line is measured along the perpendicular from the point to the line, perpendicular lines of this sort are essential in many Euclidean and hyperbolic measurements. Given two lines *l* and *m*, the distance between the lines must be measured along a line perpendicular to both *l* and *m*, guaranteeing that the distance measured in both directions is the same.

Definition 3.3.2 The *shortest distance* between two parallel lines is measured along a common perpendicular between the lines.

Theorem 3.3.8 Parallel lines cannot have more than one common perpendicular.

Let *AB* be perpendicular to *AC* and *BD* (see Figure 3.3.4). Assume that a second line *CD* is also perpendicular to both *AC* and *BD*. Then the angle

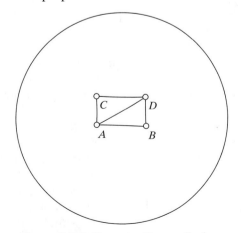

Figure 3.3.4 Common Perpendiculars

sum of quadrilateral $ABCD$ is 360°. This contradicts Theorem 3.3.4, so there is only one common perpendicular.

Theorem 3.3.9

Parallel lines are not everywhere equidistant.

A proof of this theorem is left as an exercise.

Theorem 3.3.10

Intersecting lines cannot have a common perpendicular.

A proof of this theorem is left as an exercise.

URL 3.3.1

Lambert Quadrilaterals

J. H. Lambert (1728–1777) also attempted a proof of Euclid's fifth axiom using an indirect argument. Lambert approached the proof using a quadrilateral similar to that shown in Figure 3.3.5. In Lambert quadrilateral $ABEG$, $AB \perp BE$, $BE \perp EG$, $AB \perp AG$.

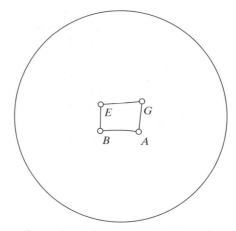

Figure 3.3.5 Lambert Quadrilateral

Theorem 3.3.11

The fourth angle of a Lambert quadrilateral is acute (see Figure 3.3.5).

A proof of this theorem is left as an exercise.

Area of Hyperbolic Triangles

In Euclidean geometry, units of area are represented in terms of unit squares. Squares are used for this purpose for a number of reasons. For instance, they tessellate the plane and fill rectangular areas in a simple and direct manner. Furthermore, there is a simple mathematical relationship between the linear dimensions of a square and its computed area. These and other features make the square a convenient shape for the unit of area in Euclidean space. However, since squares do not exist in hyperbolic space, hyperbolic area cannot be represented in terms of square units. Some other representation is needed.

In searching for a suitable unit of area for hyperbolic space, three formal considerations must be taken into account, in addition to matters of convenience. These considerations may be regarded as axioms of area.

Area Axiom 1: The area of any portion of the plane must be non-negative.

Area Axiom 2: The area of congruent portions of the plane must be the same.

Area Axiom 3: The area of the union of disjoint regions of the plane must equal the sum of the areas of the regions.

Area in the Euclidean plane (see Figure 3.3.6) may be defined as

$$A = \iint r \, dr \, d\theta$$

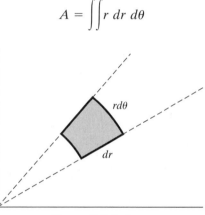

Figure 3.3.6 Area

In order to adapt this formula to the Poincare model of hyperbolic space, the "shrinking ruler" effect must be taken into account. When this is done, the result is surprising. Because demonstrating that result is beyond the scope of this text, the following theorem is stated without proof.

Theorem 3.3.12 The area of a hyperbolic triangle with degree angle sum σ is $180° - \sigma$. The area of a hyperbolic triangle with radian angle sum σ is $\pi - \sigma$.

URL 3.3.2 An interesting aspect of this result is that area is measured in degrees or radians. As a direct result of Theorem 3.3.12, if two triangles have the same angle sum, they have the same area.

Theorem 3.3.13 If two hyperbolic triangles have the same angle sum, they have the same area.

The proof of this theorem is left as an exercise.

Finally, Theorem 3.3.12 suggests that there is a largest triangle or class of largest triangles in hyperbolic space. "Largest" triangles have an angle sum of 0 and an area of 180°. A triangle with this angle sum is shown in Figure 3.3.7. It is a triangle with 3 omega (3Ω) points.

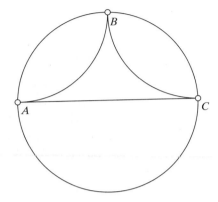

Figure 3.3.7 Area of a 3 Ω Triangle

Summary

Table 3.3.1 summarizes the definitions and theorems presented in this section.

Definitions	
3.3.1	The *defect* of a hyperbolic triangle with radian angle sum σ is $\pi - \sigma$.
3.3.2	The *distance* between two parallel lines is measured along a common perpendicular between the lines.
Theorems	
3.3.1	The summit angles of a Saccheri quadrilateral are acute.
3.3.2	The angle sum of hyperbolic triangles is less than 180°.
3.3.3	The sum of the acute angles of a right hyperbolic triangle is less than 90°.
3.3.4	The sum of the interior angles of a convex hyperbolic polygon is less than $(n - 2)180°$.
3.3.5	The angle sum of a Saccheri quadrilateral is less than 360°.
3.3.6	Rectangles do not exist in hyperbolic space.
3.3.7	In hyperbolic right triangle ABC, sides a, b, and c are opposite angles A, B, and C, respectively. Angle C is the right angle. Then $\cosh(c) = \cosh(a)\cosh(b)$.
3.3.8	Parallel lines cannot have more than one common perpendicular.
3.3.9	Parallel lines are not everywhere equidistant.
3.3.10	Intersecting lines cannot have a common perpendicular.
3.3.11	The fourth angle of a Lambert quadrilateral is acute.
3.3.12	The area of a hyperbolic triangle with angle sum σ is $\pi - \sigma$, or $180° - \sigma$.
3.3.13	If two hyperbolic triangles have the same angle sum, they have the same area.

Table 3.3.1 Summary, Section 3.3

URLs		Note: Begin each URL with the prefix http://
3.3.1	h	www-history.mcs.st-and.ac.uk/history/Mathematicians/ Lambert.html
3.3.2	s	www.geom.umn.edu/java/triangle-area/

Table 3.3.2 Section 3.3 URLs (c = concept, h = history, s = software, d = data)

Exercises

1. A hyperbolic triangle has an angle sum of 153°. Find its angular defect. Find its area. Sketch a triangle with these features.
2. A hyperbolic triangle has an area of 153°. Find its defect. Find its angle sum. Sketch a triangle with these features.
3. In hyperbolic right triangle ABC, sides a, b, and c are opposite angles A, B, and C, respectively. Angle C is the right angle. If $a = 2$ and $b = 3$, find c.
4. In hyperbolic right triangle ABC, sides a, b, and c are opposite angles A, B, and C, respectively. Angle C is the right angle. If $c = 4$ and $b = 3$, find a.
5. The figure below shows that not every set of perpendicular lines may be extended to create a Lambert quadrilateral. Sometimes, the figure doesn't "close." What considerations determine whether a given set of perpendicular lines may be used to create a Lambert quadrilateral?

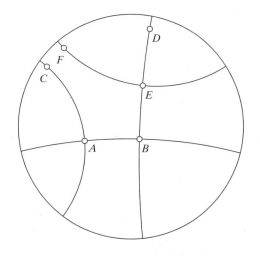

6. What range of values is possible for the sum of the interior angles of a Saccheri quadrilateral? For a Lambert quadrilateral?
7. Investigate whether the segment joining the midpoints of the sides of a Saccheri quadrilateral are perpendicular to the sides.
8. Investigate whether the following Euclidean shapes exist in hyperbolic space.
   - Parallelogram
   - Rhombus

9. In hyperbolic triangle ABC, $AC \perp BC$, $AB = 3$ and $AC = 1$. Find the length of side BC.

Prove

10. The summit of a Saccheri quadrilateral is longer than the base.
11. Cases 2 and 3 of Theorem 3.3.2.
12. The angle sum of a Saccheri quadrilateral is less than 360°.
13. Rectangles do not exist in hyperbolic space.
14. The sum of the interior angles of a convex hyperbolic polygon is less than $(n - 2)180°$.
15. Sensed-parallel lines are not everywhere equidistant.
16. Intersecting lines cannot have a common perpendicular.
17. The fourth angle of a Lambert quadrilateral is acute.
18. The sides of a Lambert quadrilateral forming the acute angle are longer than the sides opposite the acute angle.
19. Use the concept of area to prove the following statement: "If a triangle is divided into two triangles by a line intersecting one vertex and the opposite side, the defect of the original triangle is equal to the sum of the defects of the two small triangles."
20. Using an approach similar to that in #16, prove the following statement: "If a triangle is divided into two parts (one a triangle and the other a quadrilateral) by a line intersecting two sides, the defect of the original triangle is equal to the sum of the defects of the smaller triangle and quadrilateral." [Note: Assume that the defect of a quadrilateral with radian angle sum σ is computed as $2\pi - \sigma$.]
21. If three non-collinear points are located in the interior of a given hyperbolic triangle, the defect of the given triangle is greater than the defect of the triangle determined by the three interior points.

Investigation

Constructing Figures
Tool(s) Non-Euclid   *Data File(s) none*

Focus
Constructing figures in hyperbolic space

Tasks
1. Construct triangles with the following features
   - Right triangle with two omega points
   - Right triangle with one ordinary vertex and one omega point
   - Equilateral triangle
2. Construct Saccheri quadrilaterals with the following features
   - Sides the same length as the base
   - Summit angles of 45°
3. Lambert quadrilaterals with the following features
   - Acute angle of 45°
   - Congruent perpendicular sides

3.4 Congruence in Hyperbolic Space

In this section, you will . . .

- **Investigate the relationship between congruence and similarity in hyperbolic space;**
- **Explore the basis for congruence in triangles, omega triangles, and Saccheri quadrilaterals in hyperbolic space.**

Many people carry photographs of family and friends in their wallets. When these photographs are shown, it is with the clear understanding that the image in the photograph is a fair likeness of the person it depicts. Photo IDs would be useless otherwise. The process of representing a person's features on a scale small enough to fit in a wallet is a well-defined mathematical transformation, called *dilation*. Under dilation, a full-scale object and its scaled-down image are said to be *similar*. In a purely geometrical sense, two objects are similar if their corresponding angles are congruent and their corresponding sides are proportional. Common transformations like photo enlargement and reduction work because two objects may be similar in Euclidean space without being congruent. For instance, in order to carry a likeness of yourself on a photo ID, it is not necessary for the likeness to be life-sized. Photographs accurately depict mountains in 8″ × 11″ frames. And television shows us a realistic picture of the world in the corner of the living room. In hyperbolic space, however, things are different.

Theorem 3.4.1

If three angles of one triangle are congruent respectively to three angles of a second triangle, the triangles are congruent (Angle-Angle-Angle).

Let $\triangle ABC$ and $\triangle DEF$ have their corresponding angles congruent (see Figure 3.4.1). If just one pair of corresponding sides were congruent, then the triangles would be congruent by Euclid's Proposition 26. Assume that none of the three pairs of corresponding sides are congruent. Furthermore, assume that $AB > DE$ and $BC > EF$. Then we can locate D' between A and B such that $BD' \cong ED$, and F' between B and C such that $BF' \cong EF$. By Proposition 4 (SAS), $\triangle BD'F' \cong \triangle EDF$. As a result, their corresponding angles are also congruent. This forces both their defects to be the same. Since the original triangles have three pairs of congruent angles, they have identical defects. So defect ($\triangle ABC$) =

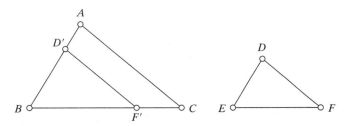

Figure 3.4.1 Congruence in Hyperbolic Space

defect ($\triangle BD'F'$). By Area Axiom 3, defect ($\triangle ABC$) = defect ($\triangle BD'F'$) + defect (quad $D'ACF'$). This forces defect (quad $D'ACF'$) = 0. If defect (quad $D'ACF'$) = 0, then its angle sum is 360°. This contradicts Theorem 3.3.4. So, $\triangle ABC \cong \triangle DEF$.

Theorem 3.4.1 makes similarity without congruence impossible in hyperbolic space. As a result, many commonplace features of life in a Euclidean universe would be impossible there. For instance, photography would only produce a realistic likeness when the image is the same size as the object. This would pose serious problems for all visually-oriented communication and information services. Images on television and movie screens would be distorted, architectural plans would have to be full-scale, and so on. Naturally, if the distortions were small, the inhabitants of hyperbolic space might not notice the distortions. When is the last time you checked?

Theorem 3.4.2

Saccheri quadrilaterals with congruent summits and summit angles are congruent.

Given Saccheri quadrilaterals $EABF$ and $SRPT$ in Figure 3.4.2, let $ST \cong EF$ and $\angle TSR \cong \angle FEA \cong \angle STP \cong \angle EFB$. Assume that the sides of the Saccheri quadrilaterals are not congruent. Without loss of generality, let $SR > EA$. Then a point J may be found on SR such that $SJ \cong EA$. A similar point I may be found on TP such that $TI \cong FB$. Let G and K be midpoints of the summits. Then $\triangle EAG \cong \triangle SJK$ and $\triangle FBG \cong \triangle TIK$ by Side-Angle-Side. Consequently $\angle EAG \cong \angle SJK$, $\angle FBG \cong \angle TIK$, $\angle EGA \cong \angle SKJ$, $\angle FGB \cong \angle TKI$, $AG \cong JK$, and $BG \cong IK$. Then $\angle AGB \cong \angle JKI$ (why?). As a result, $\triangle AGB \cong \triangle JKI$. Then $\angle GAB \cong \angle KJI$ and $\angle GBA \cong \angle KIJ$. Since $\angle EAB$ is a right angle by definition, and since $\angle EAB = \angle EAG + \angle GAB \cong \angle SJK + \angle KIJ$, $\angle SJI$ is a right angle. Then $\angle IJR$ and $\angle JIP$ are right angles (why?). As a result, quadrilateral $RPIJ$ has four right angles. Since this is impossible in hyperbolic space, the assumption that $SR > EA$ is rejected, so the Saccheri quadrilaterals are congruent.

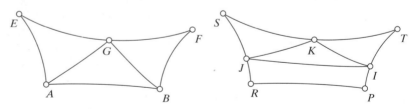

Figure 3.4.2 Congruent Saccheri Quadrilaterals

Theorem 3.4.3

Two omega triangles are congruent if the sides of finite length are congruent and if a pair of corresponding angles not located at the omega points are congruent.

Let $\triangle AE\Omega$ and $\triangle CF\Omega'$ be omega triangles with $\angle AE\Omega \cong \angle CF\Omega'$ and $AE \cong CF$ (see Figure 3.4.3). Assume that $\angle EA\Omega \neq \angle FC\Omega'$. Without

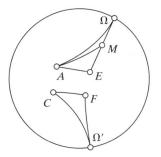

Figure 3.4.3 Congruent Omega Triangles

loss of generality, let $\angle EA\Omega > \angle FC\Omega'$. Then a point M may be located on $E\Omega$ such that $\angle EAM \cong \angle FC\Omega'$. Since M is not an omega point, $\angle AME$ is nonzero. Now $\triangle AEM \cong \triangle CF\Omega'$ by Angle-Side-Angle. Since $\angle AME$ corresponds with $\angle C\Omega'F$, they should have the same measure. But one angle is nonzero and the other is zero. This is a contradiction. So $\angle EA\Omega = \angle FC\Omega'$. Then $\triangle AE\Omega \cong \triangle CF\Omega'$ by Angle-Side-Angle.

Theorem 3.4.4   Two omega triangles $AB\Omega$ and $A'B'\Omega'$ are congruent if $\angle A \cong \angle A'$ and $\angle B \cong \angle B'$.

A proof of this theorem is left as an exercise.

Summary

It is hard to imagine life in a space in which there is no similarity without congruence. Distortion free photography, television, movies, architectural modeling, and many other applications of graphics would not be possible in hyperbolic space. This section has presented just a few examples of the way congruence differs in hyperbolic and Euclidean geometry. Table 3.4.1 summarizes the theorems presented in this section.

Theorems	
3.4.1	If three angles of one triangle are congruent respectively to three angles of a second triangle, the triangles are congruent (Angle-Angle-Angle).
3.4.2	Saccheri quadrilaterals with congruent summits and summit angles are congruent.
3.4.3	Two omega triangles are congruent if the sides of finite length are congruent and if a pair of corresponding angles not located at the omega points are congruent.
3.4.4	Two omega triangles $AB\Omega$ and $A'B'\Omega'$ are congruent if $\angle A \cong \angle A'$ and $\angle B \cong \angle B'$.

Table 3.4.1 Summary, Section 3.4

Exercises

Prove

1. Two omega triangles $AB\Omega$ and $A'B'\Omega'$ are congruent if $\angle A \cong \angle A'$ and $\angle B \cong \angle B'$.
2. Given omega triangles ACD and BCD with $AC \cong CB$, is $\triangle ACD \cong \triangle BCD$? Why or why not?

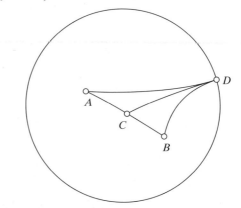

3. If the angles with ordinary vertices in an omega triangle are congruent, then the line from the omega vertex to the midpoint of the opposite side is perpendicular to that side.
4. Is $AAAA$ a congruence condition for quadrilaterals in hyperbolic space?
5. Find the flaw in the following proof:

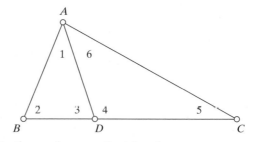

Let S denote the angle sum of a triangle.

Then
$$S = \angle 1 + \angle 2 + \angle 3 = \angle 4 + \angle 5 + \angle 6 = \angle 1 + \angle 2 + \angle 5 + \angle 6.$$

Then $2S = \angle 1 + \angle 2 + \angle 3 + \angle 4 + \angle 5 + \angle 6 = S + 180°$.
So $S = 180°$.

Investigation

Constructing Congruent Figures
Tool(s) Non-Euclid   *Data File(s) none*

Focus
Constructing congruent figures in hyperbolic space

Tasks

1. Construct pairs of congruent triangles with the following features:
   * Right triangles with two omega points
   * Right triangles with one ordinary vertex
2. Construct pairs of congruent Saccheri quadrilaterals with the following features:
   * Sides of 1 unit and bases of 2 units
   * Summit angles of 45°

References and Suggested Readings

Anderson, J. W. 1999. *Hyperbolic Geometry.* New York: Springer-Verlag.

Coxeter, H.S.M. 1988. *Non-Euclidean Geometry,* 6th ed. Washington, DC: Mathematical Association of America.

Dunham, W. 1990. *Journey Through Genius: The Great Theorems of Mathematics.* New York: Wiley, 57–60.

Finzer, W. and N. Poincare Jackiw. Disk. Available on-line http://forum. swarthmore.edu/sketchpad/gsp.gallery/poincare/poincare.html

Greenberg, M.J. 1994. *Euclidean and Non-Euclidean Geometries: Development and History,* 3rd ed. San Francisco, CA: W. H. Freeman.

Iversen, B. 1993. *An Invitation to Hyperbolic Geometry.* Cambridge, England: Cambridge University Press.

Joyce, D. *Euclid's Elements.* Available on-line http://aleph0.clarku.edu/~djoyce/java/ elements/toc.html

Lister, T. 1998. *Hyperbolic geometry using Cabri.* Available on-line http://mcs.open. ac.uk/tcl2/nonE/nonE.html

Models of the geometric plane. *The Geometry Center.* Available on-line http:// www.geom.umn.edu/docs/forum/hype/model.html

Munzner, T. 1995. *Hyperbolic Visualization.* Available on-line http://www-graphics.stanford.edu/papers/webviz/webviz/node2.html

Munzner, T. and P. Burchard. 1995. *Visualizing the structure of the WWW in 3D hyperbolic space.* Available on-line http://www-graphics. stanford.edu/papers/webviz/

O'Connor, J.J. and E.F. Robertson. 1999. *The MacTutor History of Mathematics archive.* School of Mathematics and Statistics, University of St. Andrews, Scotland.

   Bolyai.   http://www-history.mcs.st-and.ac.uk/history/Mathematicians/ Bolyai.html

   Einstein.   http://www-history.mcs.st-and.ac.uk/history/Mathematicians/ Einstein.html

   Euclid of Alexandria.   http://www-history.mcs.st-and.ac.uk/history/ Mathematicians Euclid.html

   Lobachevsky.   http://www-history.mcs.st-and.ac.uk/history/ Mathematicians/Lobachevsky.html

   Poincare.   http://www-history.mcs.st-and.ac.uk/history/Mathematicians/ Poincare.html

   Proclus Diadocus.   http://www-history.mcs.st-and.ac.uk/history/ Mathematicians/Proclus.html

   Saccheri.   http://www-history.mcs.st-and.ac.uk/history/Mathematicians/ Saccheri.html

Ramsay, A. and R.D. Richtmeyer. 1995. *Introduction to Hyperbolic Geometry.* New York: Springer-Verlag.

Robles, C. 1996. *The world of hyperbolic geometry.* Available on-line
   http://www.math. ubc.ca/~robles/hyperbolic/index.html
Rosien, A. 1997. *The area of triangles in hyperbolic geometry.* The Geometry
   Center. Available on-line   http://www.geom.umn.edu/java/triangle-area/
Royster, D. 1997. *Hyperbolic Geometry Links.* Available on-line
   http://www.math. uncc.edu/~droyster/courses/fall96/math3181/hypgeom.html
Sobral. 1919. *Eclipses at Greenwich.* Royal Observatory. Available on-line
    http://www.rog.nmm.ac.uk/astroweb/eclipses/greenwich/1919/index.html
Stillwell, J. 1996. *Sources of Hyperbolic Geometry.* Providence, RI: American
   Mathematical Society.
Sved, M. 1991. *Journey into Geometries.* Washington, DC: Mathematical
   Association of America.
Trudeau, R.J. 1987. *The Non-Euclidean Revolution.* Boston, MA: Birkhäuser.
Wells, D. 1991. *The Penguin Dictionary of Curious and Interesting Geometry.*
   London: Penguin, 109–110.

Chapter 4

Transformation Geometry

URLs
4.1.1
–
4.1.5

From the time of Euclid (300 B.C.) to the 17th century, geometry was studied entirely from a synthetic perspective. During the 17th century a number of newly developed mathematical ideas were applied to the study of geometry, with revolutionary effects. For instance, in applying algebraic concepts and notations to geometry, Pierre de Fermat (1601–1665) and Rene Descartes (1596–1650) created analytic geometry. Differential geometry evolved as concepts and notations from calculus developed by Newton and Leibniz were applied to the study of geometry. During the 18th and 19th centuries, a number of non-Eudlidean geometries were developed, causing some to wonder if geometry itself was splitting apart into competing theories. In 1872, a 23-year-old mathematician named Felix Klein (1849–1925) proposed a new unifying principle for classifying various geometries and explaining the relationships between them. At the heart of Klein's principle is the concept of a geometric transformation.

Geometric transformations are one-to-one mappings, taking point sets as inputs and returning point sets as outputs. For simplicity, input sets are called *objects* and their corresponding output sets are called *images.* Depending on the context, transformations may be conceived of as applying to familiar geometric objects such as lines, polygons, and polyhedra or to the spaces in which they are embedded.

Transformation geometry offers deep insights into the nature of many traditional topics, including congruence, similarity, and symmetry. It is also the basis for many contemporary applications in the arts, architecture, engineering, film and television. A deeper significance, however, is seen in Felix Klein's definition of a geometry: "A geometry is the study of those properties of a set S that remain invariant (unchanged) when the elements of S are subjected to the transformations of some transformation group." This definition, given in Klein's Erlanger Program of 1872, establishes transformation geometry as a means of understanding the relationships among all geometries, Euclidean and non-Euclidean. This chapter focuses on geometric transformations of the Euclidean plane, their matrix representations, invariant features, and applications.

4.1 An Analytic Model of the Euclidean Plane

In this section, you will . . .

Use matrix notation to:
- **Represent points and lines in the Euclidean plane;**
- **Find the area of a triangle given its vertices;**
- **Determine the equation of a line given two points on the line; and**
- **Determine the point of intersection of two lines.**

Representing Points and Lines

In order to discuss an analytic model of the Euclidean plane, some plane in 3-space must be selected to serve in that capacity. To many students, the x-y plane, $z = 0$, seems a natural choice. There is a better choice, however, the plane $z = 1$. While the benefits of this choice are immediately apparent in the convenient notations presented in this chapter, a broader discussion of the merits of this choice is deferred until the chapter on projective geometry provides a suitable motivation and mathematical context.

Figure 4.1.1 shows the plane $z = 1$ in a standard right-handed coordinate system. Every point in this plane has coordinates of the form $(x, y, 1)$.

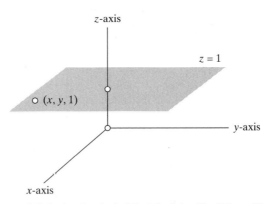

Figure 4.1.1 An Analytic Model of the Euclidean Plane

In most introductory algebra courses, lines in the Euclidean plane are represented using equations in slope-intercept $y = mx + b$ or general form $ax + by = c$. While these representations retain their utility in many contexts, a more elegant, versatile, and powerful representation may be developed using matrix notation. Matrix notation was first introduced by Arthur Cayley (1821–1895). One of the principal goals of this chapter is to introduce matrix representations for points, lines and their relationships.

URL
4.1.6

In matrix notation, the coefficients and constants in algebraic equations of the form $u_1 x + u_2 y + u_3 = 0$ are written as row vectors

$$[u_1 \quad u_2 \quad u_3]$$

Points are written as column vectors

$$\begin{bmatrix} x \\ y \\ 1 \end{bmatrix}$$

Combining these two notations, we obtain the matrix equation

$$\begin{bmatrix} u_1 & u_2 & u_3 \end{bmatrix} \begin{bmatrix} x \\ y \\ 1 \end{bmatrix} = 0$$

When corresponding row and column entries are multiplied and the resulting terms added, an algebraic equation is obtained. For example, the matrix equation

$$\begin{bmatrix} 1 & 3 & 2 \end{bmatrix} \begin{bmatrix} x \\ y \\ 1 \end{bmatrix} = 0$$

is equivalent to the algebraic equation $1x + 3y + 2 = 0$. Equations of this sort are often used to answer the question, "Which points in the plane are on the given line?"

Area and Collinearity

Matrix notation provides a compact and easily remembered format for many equations in analytic geometry. The following example illustrates this fact while addressing the question, "What area is determined by three noncollinear points?"

Example 4.1.1   Any three noncollinear points in the plane determine a triangle (see Figure 4.1.2). Given the coordinates of the triangle's vertices, find the area of the triangle.

CD 4.1.1

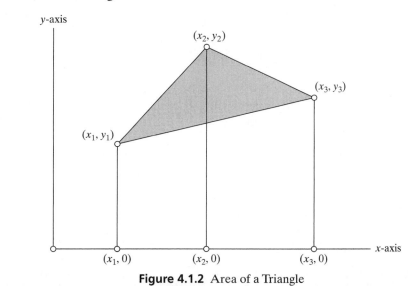

Figure 4.1.2 Area of a Triangle

The area of the triangle may be written in terms of the areas of three trapezoids.

Trapezoid 1 Vertices: $(x_1, 0, 1), (x_1, y_1, 1), (x_2, y_2, 1), (x_2, 0, 1)$
Trapezoid 2 Vertices: $(x_2, 0, 1), (x_2, y_2, 1), (x_3, y_3, 1), (x_3, 0, 1)$
Trapezoid 3 Vertices: $(x_1, 0, 1), (x_1, y_1, 1), (x_3, y_3, 1), (x_3, 0, 1)$

Using these representations for the areas of the trapezpoids, the area of the triangle is given by

$$[\text{Area of Trapezoid 1}] + [\text{Area of Trapezoid 2}] - [\text{Area of Trapezoid 3}]$$

$$= \frac{1}{2}[(x_2 - x_1)(y_1 + y_2) + (x_3 - x_2)(y_2 + y_3) - (x_3 - x_1)(y_1 + y_3)]$$

Expanding, simplifying, and arranging terms yields the expression

$$\frac{1}{2}[x_2 y_1 - x_1 y_2 + x_3 y_2 - x_2 y_3 - x_3 y_1 + x_1 y_3]$$

Using a matrix approach, the coordinates of the triangle's vertices may be written as column vectors in the matrix

$$\begin{bmatrix} x_1 & x_2 & x_3 \\ y_1 & y_2 & y_3 \\ 1 & 1 & 1 \end{bmatrix}$$

The determinant of this matrix is written as

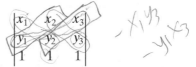

Expanding this determinant yields the expression

$$[x_1 y_2 + x_3 y_1 + x_2 y_3 - x_3 y_2 - x_2 y_1 - x_1 y_3]$$

which may be rearranged as

$$[-x_2 y_1 + x_1 y_2 - x_3 y_2 + x_2 y_3 + x_3 y_1 - x_1 y_3]$$

Comparing this to the expression obtained for the area of the triangle, it follows that the area of the triangle may be written as:

$$\text{Area} = \left(\frac{1}{2}\right)\text{abs}\left(\begin{vmatrix} x_1 & x_2 & x_3 \\ y_1 & y_2 & y_3 \\ 1 & 1 & 1 \end{vmatrix}\right)$$

This equation is both easy to remember and easy to evaluate, assuming that the user knows how to compute the determinant of a 3×3 matrix. The result is formalized in Theorem 4.1.1.

Theorem 4.1.1   Given any three noncollinear points in the plane, the area of the triangle determined by three points is given by the formula

$$\text{Area} = \left(\frac{1}{2}\right)\text{abs}\left(\begin{vmatrix} x_1 & x_2 & x_3 \\ y_1 & y_2 & y_3 \\ 1 & 1 & 1 \end{vmatrix}\right)$$

A proof of this theorem, based on Example 4.1.1, is left as an exercise.

The Equation of a Line

Figure 4.1.2 and Theorem 4.1.1 provide two valuable insights into a related problem, finding an equation of the line through two given points in the Euclidean plane. First, Figure 4.1.2 suggests that the triangular area associated with three collinear points should be zero. Second, Theorem 4.1.1 suggests that, under such circumstances, the determinant

$$\begin{vmatrix} x_1 & x_2 & x_3 \\ y_1 & y_2 & y_3 \\ 1 & 1 & 1 \end{vmatrix}$$

should also equal zero. It follows that, given two points

$$\begin{bmatrix} x_1 \\ y_1 \\ 1 \end{bmatrix} \quad \text{and} \quad \begin{bmatrix} x_2 \\ y_2 \\ 1 \end{bmatrix}, \quad \text{all additional points} \quad \begin{bmatrix} x \\ y \\ 1 \end{bmatrix}$$

collinear with the given points must satisfy the matrix equation

$$\begin{vmatrix} x & x_1 & x_2 \\ y & y_1 & y_2 \\ 1 & 1 & 1 \end{vmatrix} = 0$$

Example 4.1.2   Find an equation of the line containing the points (2, 4, 1) and (3, 1, 1). Writing the given points as column vectors, and adding a third column to represent all points collinear with the given points, we obtain the matrix equation

$$\begin{vmatrix} x & 2 & 3 \\ y & 4 & 1 \\ 1 & 1 & 1 \end{vmatrix} = 0$$

Expanding and simplifying, we obtain

$$\begin{vmatrix} x & 2 & 3 \\ y & 4 & 1 \\ 1 & 1 & 1 \end{vmatrix} = 4x + 3y + 2 - x - 2y - 12 \Rightarrow 3x + y - 10 = 0, \quad \text{or}$$

$$1\begin{vmatrix} 2 & 3 \\ 4 & 1 \end{vmatrix} - 1\begin{vmatrix} x & 3 \\ y & 1 \end{vmatrix} + 1\begin{vmatrix} x & 2 \\ y & 4 \end{vmatrix}$$

$$[3 \quad 1 \quad -10]\begin{bmatrix} x \\ y \\ 1 \end{bmatrix} = 0.$$

$$2 + 12 - x - 3y + 4x - 2y = 0$$

$$3x - 5y + 14$$

The elegance and power of this approach to simplify both the representation and solution of such problems is readily apparent. The result is formalized in Theorem 4.1.2.

Theorem 4.1.2

Given two points

$$\begin{bmatrix} x_1 \\ y_1 \\ 1 \end{bmatrix} \quad \text{and} \quad \begin{bmatrix} x_2 \\ y_2 \\ 1 \end{bmatrix}$$

an equation of the line containing the points is determined by the matrix equation

$$\begin{vmatrix} x & x_1 & x_2 \\ y & y_1 & y_2 \\ 1 & 1 & 1 \end{vmatrix} = 0$$

A proof of this theorem, based on Example 4.1.2, is left as an exercise.

The Intersection of Two Lines

Matrix methods may also be used to find the point of intersection of two lines in the plane. Given the equations $u_1 x + u_2 y + u_3 = 0$ and $v_1 x + v_2 y + v_3 = 0$, the coordinates of the point of intersection may be determined using either linear combination or substitution then expressed in matrix notation as follows:

$$x = \frac{(u_2 v_3 - u_3 v_2)}{(u_1 v_2 - u_2 v_1)} = \frac{\begin{vmatrix} u_2 & u_3 \\ v_2 & v_3 \end{vmatrix}}{\begin{vmatrix} u_1 & u_2 \\ v_1 & v_2 \end{vmatrix}} \quad \text{and} \quad y = \frac{(u_3 v_1 - u_1 v_3)}{(u_1 v_2 - u_2 v_1)} = \frac{\begin{vmatrix} u_3 & u_1 \\ v_3 & v_1 \end{vmatrix}}{\begin{vmatrix} u_1 & u_2 \\ v_1 & v_2 \end{vmatrix}}$$

An obvious shortcoming of this approach is that two equations must be remembered and properly evaluated. A better approach would require only one equation to provide both coordinates of the point of intersection. Writing the line parameters as row elements in the matrix equation

$$\begin{vmatrix} u_1 & u_2 & u_3 \\ v_1 & v_2 & v_3 \\ a & b & c \end{vmatrix} = 0$$

provides a simpler representation. Expanding by the last row, this equation may be rewritten as

$$a \begin{vmatrix} u_2 & u_3 \\ v_2 & v_3 \end{vmatrix} - b \begin{vmatrix} u_1 & u_3 \\ v_1 & v_3 \end{vmatrix} + c \begin{vmatrix} u_1 & u_2 \\ v_1 & v_2 \end{vmatrix} = 0$$

Dividing by the 2×2 matrix in the last term yields the equation

$$a\frac{\begin{vmatrix} u_2 & u_3 \\ v_2 & v_3 \end{vmatrix}}{\begin{vmatrix} u_1 & u_2 \\ v_1 & v_2 \end{vmatrix}} - b\frac{\begin{vmatrix} u_1 & u_3 \\ v_1 & v_3 \end{vmatrix}}{\begin{vmatrix} u_1 & u_2 \\ v_1 & v_2 \end{vmatrix}} + c\frac{\begin{vmatrix} u_1 & u_2 \\ v_1 & v_2 \end{vmatrix}}{\begin{vmatrix} u_1 & u_2 \\ v_1 & v_2 \end{vmatrix}} = 0$$

Recalling that

$$x = \frac{\begin{vmatrix} u_2 & u_3 \\ v_2 & v_3 \end{vmatrix}}{\begin{vmatrix} u_1 & u_2 \\ v_1 & v_2 \end{vmatrix}} \quad \text{and} \quad y = \frac{\begin{vmatrix} u_3 & u_1 \\ v_3 & v_1 \end{vmatrix}}{\begin{vmatrix} u_1 & u_2 \\ v_1 & v_2 \end{vmatrix}}$$

this equation may be simplified to $ax + by + c = 0$ and written in matrix form as

$$\begin{bmatrix} a & b & c \end{bmatrix} \begin{bmatrix} x \\ y \\ 1 \end{bmatrix} = 0$$

In this case, this is more than the general matrix form for the equation of a line because, while the line parameters $\begin{bmatrix} a & b & c \end{bmatrix}$ are arbitrary, x and y are the coordinates of the point of intersection of the given lines $u_1x + u_2y + u_3 = 0$ and $v_1x + v_2y + v_3 = 0$. This result is formalized in Theorem 4.1.3 and illustrated in Example 4.1.3.

Theorem 4.1.3 Given the equations $u_1x + u_2y + u_3 = 0$ and $v_1x + v_2y + v_3 = 0$, the point of intersection of u and v

$$\begin{bmatrix} x \\ y \\ 1 \end{bmatrix}$$

is given by the equation

$$\begin{bmatrix} a & b & c \end{bmatrix} \begin{bmatrix} x \\ y \\ 1 \end{bmatrix} = 0$$

obtained by expanding and simplifying the equation

$$\begin{vmatrix} u_1 & u_2 & u_3 \\ v_1 & v_2 & v_3 \\ a & b & c \end{vmatrix} = 0$$

A proof of this theorem is left as an exercise.

Example 4.1.3   Find the coordinates of the point of intersection of $1x + 1y + 1 = 0$ and $1x - 1y + 2 = 0$. The matrix equation is

$$\begin{vmatrix} 1 & 1 & 1 \\ 1 & -1 & 2 \\ a & b & c \end{vmatrix} = 0$$

Expanding and simplifying yields

$$\begin{vmatrix} 1 & 1 & 1 \\ 1 & -1 & 2 \\ a & b & c \end{vmatrix} = 3a - b - 2c = 0$$

This may be rewritten as

$$\begin{bmatrix} a & b & c \end{bmatrix} \begin{bmatrix} 3 \\ -1 \\ -2 \end{bmatrix} = 0$$

Since the column vector

$$\begin{bmatrix} 3 \\ -1 \\ -2 \end{bmatrix}$$

is not a point of the Euclidean plane, each element in the vector is divided by the last entry, -2, producing the Euclidean point

$$\begin{bmatrix} -\dfrac{3}{2} \\ \dfrac{1}{2} \\ 1 \end{bmatrix}$$

We are now ready to answer the question, "Given the lines $1x + 1y + 1 = 0$ and $1x - 1y + 2 = 0$, what is their point of intersection?" The answer is the point

$$\begin{bmatrix} -\dfrac{3}{2} \\ \dfrac{1}{2} \\ 1 \end{bmatrix}$$

In addition, all lines on that point must satisfy the same matrix equation.

$$\begin{bmatrix} a & b & c \end{bmatrix} \begin{bmatrix} -\dfrac{3}{2} \\ \dfrac{1}{2} \\ 1 \end{bmatrix} = 0$$

So, while $1x + 1y + 1 = 0$ and $1x + 1y + 2 = 0$ are on the point, so also are all other lines $[a \quad b \quad c]$ satisfying the same matrix equation.

An important distinction may now be made between two uses of notations of the form

$$[a \quad b \quad c] \begin{bmatrix} x \\ y \\ 1 \end{bmatrix} = 0$$

1. When the values represented by the row vector $[a \quad b \quad c]$ are known and the values represented by the column vector

$$\begin{bmatrix} x \\ y \\ 1 \end{bmatrix}$$

are unknown, such as

$$[1 \quad -2 \quad 5] \begin{bmatrix} x \\ y \\ 1 \end{bmatrix} = 0,$$

the notation is called "an equation of the line" and is normally used to answer the question, "Which points are on this line?"
2. When the situation is reversed and the values represented by the row vector $[a \quad b \quad c]$ are unknown while the values represented by the column vector

$$\begin{bmatrix} x \\ y \\ 1 \end{bmatrix}$$

are known, such as

$$[a \quad b \quad c] \begin{bmatrix} 1 \\ 3 \\ 1 \end{bmatrix} = 0,$$

the notation is called "an equation of the point" and is normally used to answer the question, "Which lines are on this point?" This symmetry of expression, achieved by interchanging the terms *point* and *line,* is known as *duality* and is relatively rare in Euclidean geometry.

Summary

The first step in creating an analytic geometry of the Euclidean plane is to decide which plane to use, among infinitely many possibilities, and how to represent points, lines, and their relationships. Matrix notation has been introduced because of its concise, uncluttered format and in order to take advantage of many powerful results from matrix algebra that apply to the study of geometry. The notations introduced in this section are summarized below.

Representing Points of the Euclidean Plane

Q: Where is the following point located?
A: The coordinates are $(x, y, 1)$.

Lines in the Euclidean Plane

Q: Where is the following line located?
A: It crosses the y-axis at the point $(0, -c/b, 1)$ and has slope $(-a/b)$.

$$\begin{bmatrix} a & b & c \end{bmatrix}$$

Incidence of a Point and a Line

Q: Is the point $(x, y, 1)$ on the line $\begin{bmatrix} a & b & c \end{bmatrix}$?
A: If $ax + by + c = 0$, the point is on the line.

$$\begin{bmatrix} a & b & c \end{bmatrix}\begin{bmatrix} x \\ y \\ 1 \end{bmatrix} = 0$$

The Equation of a Line

Q: What points are on the line $\begin{bmatrix} a & b & c \end{bmatrix}$?
A: Every point $(x, y, 1)$ that satisfies the equation $ax + by + c = 0$.

$$\begin{bmatrix} a & b & c \end{bmatrix}\begin{bmatrix} x \\ y \\ 1 \end{bmatrix} = 0$$

For example,

$$\begin{bmatrix} 1 & 3 & -4 \end{bmatrix}\begin{bmatrix} x \\ y \\ 1 \end{bmatrix} = 0$$

The Equation of a Point

Q: What lines are on the point $(x, y, 1)$?
A: Every line $\begin{bmatrix} a & b & c \end{bmatrix}$ that satisfies the equation $ax + by + c = 0$.

$$\begin{bmatrix} a & b & c \end{bmatrix}\begin{bmatrix} x \\ y \\ 1 \end{bmatrix} = 0$$

For example,

$$[a \ \ b \ \ c] \begin{bmatrix} 1 \\ 2 \\ 1 \end{bmatrix} = 0$$

The Area of a Triangle

Q: Given three points $(x_1, y_1, 1), (x_2, y_2, 1), (x_3, y_3, 1)$ what is the area of the triangle determined by the points?

A: Writing the points as column vectors in a matrix, the area is half the absolute value of the determinant.

$$\text{Area} = \left(\frac{1}{2}\right) \text{abs} \left(\begin{vmatrix} x_1 & x_2 & x_3 \\ y_1 & y_2 & y_3 \\ 1 & 1 & 1 \end{vmatrix} \right)$$

The Equation of a Line

Q: Given two points $(p, q, 1)$ and $(r, s, 1)$, what is the equation of the line containing the points?

A: Writing the points as column vectors in a matrix, and inserting a third column to represent other collinear points $(x, y, 1)$, the equation of the line is determined by taking the determinant of the matrix.

$$\begin{vmatrix} x & p & r \\ y & q & s \\ 1 & 1 & 1 \end{vmatrix} = 0$$

The Equation of a Point

Q: Given two lines $u_1x + u_2y + u_3 = 0$ and $v_1x + v_2y + v_3 = 0$, what are the coordinates of the point of intersection, $(x, y, 1)$?

A: Writing the given lines as row vectors in a matrix, and inserting a third row to represent other concurrent lines $[a \ \ b \ \ c]$, the equation of the point

is determined by taking the determinant of the matrix

$$\begin{vmatrix} u_1 & u_2 & u_3 \\ v_1 & v_2 & v_3 \\ a & b & c \end{vmatrix} = 0$$

and writing the resulting expression in the form

$$[a \ \ b \ \ c] \begin{bmatrix} x \\ y \\ 1 \end{bmatrix} = 0$$

URLs		Note: Begin each URL with the prefix http://
4.1.1	h	http://www-history.mcs.st-andrews.ac.uk/history/ Mathematicians/Fermat.html
4.1.2	h	http://www-history.mcs.st-andrews.ac.uk/history/ Mathematicians/Descartes.html
4.1.3	h	http://www-history.mcs.st-andrews.ac.uk/history/ Mathematicians/Newton.html
4.1.4	h	http://www-history.mcs.st-andrews.ac.uk/history/ Mathematicians/Leibniz.html
4.1.5	h	http://www-history.mcs.st-andrews.ac.uk/history/ Mathematicians/Klein.html
4.1.6	h	http://www-history.mcs.st-andrews.ac.uk/history/ Mathematicians/Cayley.html

Table 4.1.1 Section 4.1 URLs (c = concept, h = =history, s = software, d = data)

Exercises

1. Write equations of the following lines using matrix notation
   a) $x + y = 0$
   b) $2x + 3y - 5 = 0$
   c) $y = 3x - 1$
   d) $x = 4$
   e) $y = -3$
2. Using the following points as vertices of a triangle and matrix methods, find the area of the triangle.
   a) $(0, 0) (1, 0) (0, 1)$
   b) $(0, 0) (2, 0) (2, 2)$
   c) $(-1, 1) (0, -1) (8, 0)$
   d) $(1, 1) (2, -3) (-6, 4)$
3. Using matrix methods, find a matrix equation of the line through the given points.
   a) $(0, 0) (1, 0)$
   b) $(1, 1) (1, 2)$
   c) $(1, 1) (2, -3)$
   d) $(-1, 1) (0, -1)$
4. Using matrix methods, find the point of intersection of the following lines.
   a) $[2 \quad -1 \quad 5]$ and $[-3 \quad 2 \quad 1]$
   b) $y = 2x + 1$ and $x - 4y + 1 = 0$
   c) $[1 \quad 1 \quad 0]$ and $[0 \quad 2 \quad -1]$
   d) $[0 \quad 1 \quad 2]$ and $[1 \quad -1 \quad 0]$
5. As a high school geometry teacher, how would you implement the NCTM recommendations regarding the introduction of matrix methods in the study of transformation geometry?

$-y = x + 6$
$x - y = 6$
$-x - y = -16$
$-1 - 16$

$-1 \quad 1 \quad 6$

$1 \quad 0 \quad 2$

$y = 2$

6. Using the *MacTutor History of Mathematics Archive* on the WWW, write a brief summary of the role that transformation geometry played in Klein's approach to developing a synthesis of many geometries.
7. Prove Theorem 4.1.1.
8. Prove Theorem 4.1.2.
9. Prove Theorem 4.1.3.
10. Under what conditions will a line $[u_1 \quad u_2 \quad u_3]$ not intersect a line $[v_1 \quad v_2 \quad v_3]$?
11. Under what conditions will a line $[u_1 \quad u_2 \quad u_3]$ be perpendicular to a line $[v_1 \quad v_2 \quad v_3]$?
12. Under what conditions will a line $[u_1 \quad u_2 \quad u_3]$ pass through the origin and be parallel to the line $[v_1 \quad v_2 \quad v_3]$?

4.2 Representing Linear Transformations in 2-Space with Matrices

In this section, you will . . .

- Begin investigating the geometric features of a class of linear transformations known as isometries;
- Associate the concept of a linear transformation with the concept of a function;
- Represent linear transformations using matrix notation; and
- Find the images of points and lines under linear transformations using matrix methods.

Isometries

Of all geometric transformations, the isometries are most basic. The word *isometry* contains within its structure the key to its meaning, *isometry* denoting *same measurement*. When an isometry is applied to an object, the object and its image have the same linear and angular measurements. The transformation is said to *preserve* these features, and the features are said to be *invariant* under the transformation. Preserving linear and angular measurements guarantees that perimeter, angle-sum, and area are also preserved. As a result, the object and its image under an isometry are identical, or congruent.

Many everyday acts, such as sliding a book on a desktop, rotating a plate on a table, and laying down footprints in the sand, may be viewed as isometries. Such actions reposition objects without changing their geometric characteristics. Because most movements in the real world preserve the geometric characteristics of objects in motion (car crashes excepted), an understanding of the mathematics of isometries is fundamental to representing motion and/or changes in perspective in computer simulations and other modeling environments. The National Council of Teachers of Mathematics recommends that all students learn to recognize and use geometric transformations in the study of mathematics and its applications in science and other pursuits.

URL 4.2.1

CD 4.2.1 – 4.2.3

Isometries of the Euclidean plane fall into three categories and their compositions: translations (slides), rotations (turns), and reflections (flips). Of all isometries, translations are the simplest to understand. Under the operations of translation, every point in an object moves the same distance parallel to a slide arrow, or vector (see Figure 4.2.1 and cd4_2_1). Under the operation of rotation, every point is moved through a turn angle relative to a turn center (see Figure 4.2.2 and cd4_2_2). Reflections map every point across a reflecting line a distance equal to its distance from the reflecting line (see Figure 4.2.3 and cd4_2_3).

Definition 4.2.1 Let A and B be sets. A *mapping or function f* from A to B is a rule that assigns to each $x \in A$ exactly one $y \in B$ and is written $y = f(x)$.

Definition 4.2.2 A mapping from A to B is *onto* if, for every $y \in B$, there is at least one $x \in A$ such that $y = f(x)$.

**URL
4.2.2**

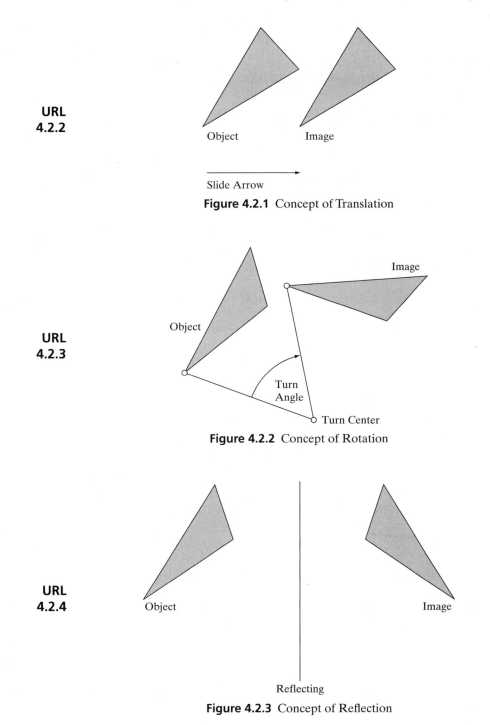

Object   Image

Slide Arrow

Figure 4.2.1 Concept of Translation

**URL
4.2.3**

Image

Object

Turn
Angle

Turn Center

Figure 4.2.2 Concept of Rotation

**URL
4.2.4**

Object   Image

Reflecting

Figure 4.2.3 Concept of Reflection

**URL
4.2.5**

A number of definitions are introduced now to facilitate a careful investigation of linear transformations and their properties as they apply to the Cartesian coordinate plane.

Definition 4.2.3 A mapping f from A to B is *one-to-one* if whenever $x \neq y$, $f(x) \neq f(y)$.

Definition 4.2.4 If $f: A \to B$ is a one-to-one, onto mapping, then $f^{-1}: B \to A$ is the *inverse* of f if $(f^{-1}*f)(x) = x$ for all $x \in A$ and $(f*f^{-1})(y) = y$ for all $y \in B$.

Definition 4.2.5 The Euclidean plane consists of the set of all points X such that $X = (x_1, x_2, 1)$, where x_i belongs to the set of real numbers, written $x_i \in \mathbb{R}$.

Definition 4.2.6 Let V be a vector space over $\mathbb{R}$ and T a function from V to V. T is a linear transformation of V if $T(u + v) = T(u) + T(v)$ for all vectors $u \in V$ and $v \in V$ and $T(ku) = kT(u)$ for all vectors $u \in V$ and scalars $k \in \mathbb{R}$.

Representing Linear Transformations with Matrices

Definition 4.2.7 An *invertible linear transformation* $T(X)$ of the Euclidean plane is a one-to-one, onto mapping of points from the Euclidean plane onto the Euclidean plane. $T(X) = A*X$, where $|A| \neq 0$ and

$$A = \begin{bmatrix} a_{11} & a_{12} & a_{13} \\ a_{21} & a_{22} & a_{23} \\ 0 & 0 & 1 \end{bmatrix}, \quad \text{where } a_{ij} \in \mathbb{R}$$

Under a linear transformation, each point in the image plane is associated with a single point in the object plane. No points are omitted. None disappear. There is a complete Euclidean plane before and after the transformation. For instance, Figure 4.2.4 shows a rectangular coordinate system in the Euclidean plane before and after a translation. Naturally, when a linear transformation of this sort is applied to the plane, all objects embedded in the plane undergo the same transformation.

Linear transformations may be composed, one after the other, in a sequence. For example a translation T may be followed by a rotation R. In functional notation, this is written $R(T(X))$, where X is some object set. In matrix notation, the same composition is written

$$RT \begin{bmatrix} x \\ y \\ 1 \end{bmatrix}$$

Theorem 4.2.1 The composition of two linear transformations of the plane is itself a linear transformation of the plane.

According to Definition 4.2.6, T is a linear transformation of the plane if $T(u + v) = T(u) + T(v)$ and $T(ku) = kT(u)$ for all points

$$u = \begin{bmatrix} u_1 \\ u_2 \\ 1 \end{bmatrix}, \quad v = \begin{bmatrix} v_1 \\ v_2 \\ 1 \end{bmatrix} \quad \text{and real numbers } k.$$

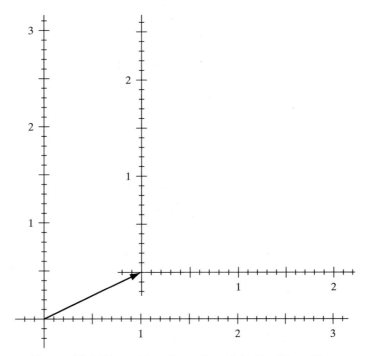

Figure 4.2.4 Linear Transformation of the Euclidean Plane

Define two linear transformations

$$T_1 = \begin{bmatrix} a_1 & b_1 & e_1 \\ c_1 & d_1 & f_1 \\ 0 & 0 & 1 \end{bmatrix} \quad \text{and} \quad T_2 = \begin{bmatrix} a_2 & b_2 & e_2 \\ c_2 & d_2 & f_2 \\ 0 & 0 & 1 \end{bmatrix}$$

The composition $T_1 T_2$ may be computed as

$$\begin{bmatrix} a_1 & b_1 & e_1 \\ c_1 & d_1 & f_1 \\ 0 & 0 & 1 \end{bmatrix} \begin{bmatrix} a_2 & b_2 & e_2 \\ c_2 & d_2 & f_2 \\ 0 & 0 & 1 \end{bmatrix} \begin{bmatrix} x \\ y \\ 1 \end{bmatrix}$$

$$= \begin{bmatrix} (a_1 a_2 + b_1 c_2) & (a_1 b_2 + b_1 d_2) & (a_1 e_2 + b_1 f_2 + e_1) \\ (c_1 a_2 + d_1 c_2) & (c_1 b_2 + d_1 d_2) & (c_1 e_2 + d_1 f_2 + f_1) \\ 0 & 0 & 1 \end{bmatrix}$$

Each indicated product and sum is itself a real number, no two of which have identical representations. Because addition and multiplication are closed operations in the real numbers, each expression yields a real number when evaluated. So, the composition

$$T_1 T_2 \begin{bmatrix} x \\ y \\ 1 \end{bmatrix}$$

may be rewritten as

$$\begin{bmatrix} a & b & e \\ c & d & f \\ 0 & 0 & 1 \end{bmatrix} \begin{bmatrix} x \\ y \\ 1 \end{bmatrix}, \text{ where } a, b, c, d, e, f \in R$$

If we let

$$A = \begin{bmatrix} a & b & e \\ c & d & f \\ 0 & 0 & 1 \end{bmatrix}$$

the only remaining condition to be satisfied is that $|A| \neq 0$. Verifying this fact involves considerable symbol manipulation.

Experience with linear transformations in the real world suggests that at least some linear transformations are reversible. For instance, one may translate or rotate an object, then return it to its starting position. The mathematical equivalent of this notion of reversibility is that of the inverse of a function f, written f^{-1}.

If $A = \begin{bmatrix} a & b & e \\ c & d & f \\ 0 & 0 & 1 \end{bmatrix}$ is a linear transformation, A^{-1} is written $\begin{bmatrix} a & b & e \\ c & d & f \\ 0 & 0 & 1 \end{bmatrix}^{-1}$.

Theorem 4.2.2   Given a linear transformation T, there exists a corresponding inverse transformation T^{-1} such that $TT^{-1} = T^{-1}T = I$, where I is the identity transformation.

From linear algebra, we know that a matrix A has an inverse if $|A| \neq 0$.

The Image of a Point Under a Linear Transformation

Using matrix notation, the image of a point under a given linear transformation is defined as follows.

Definition 4.2.8   Given a linear transformation

$$A = \begin{bmatrix} a & b & e \\ c & d & f \\ 0 & 0 & 1 \end{bmatrix}$$

the image of point

$$X = \begin{bmatrix} x \\ y \\ 1 \end{bmatrix}$$

under transformation A is given by

$$\begin{bmatrix} a & b & e \\ c & d & f \\ 0 & 0 & 1 \end{bmatrix} \begin{bmatrix} x \\ y \\ 1 \end{bmatrix} = \begin{bmatrix} ax + by + e \\ cx + dy + f \\ 1 \end{bmatrix} = \begin{bmatrix} x' \\ y' \\ 1 \end{bmatrix}$$

Example 4.2.1   The matrix equation

$$\begin{bmatrix} a & b & e \\ c & d & f \\ 0 & 0 & 1 \end{bmatrix} \begin{bmatrix} x \\ y \\ 1 \end{bmatrix} = \begin{bmatrix} x' \\ y' \\ 1 \end{bmatrix}$$

is used to find the image of a point under a linear transformation. This equation is also written $AX = X'$. For instance, the image of the point

$$\begin{bmatrix} 1 \\ 3 \\ 1 \end{bmatrix}$$

under the linear transformation

$$A = \begin{bmatrix} 1 & 0 & 1 \\ 0 & 1 & 2 \\ 0 & 0 & 1 \end{bmatrix}$$

URL 4.2.6   is computed as $\begin{bmatrix} 1 & 0 & 1 \\ 0 & 1 & 2 \\ 0 & 0 & 1 \end{bmatrix} \begin{bmatrix} 1 \\ 3 \\ 1 \end{bmatrix} = \begin{bmatrix} 2 \\ 5 \\ 1 \end{bmatrix}$ and may be written as

$$A \begin{bmatrix} 1 \\ 3 \\ 1 \end{bmatrix} = \begin{bmatrix} 2 \\ 5 \\ 1 \end{bmatrix}.$$

Theorem 4.2.3   Given a linear transformation

$$T = \begin{bmatrix} a & b & e \\ c & d & f \\ 0 & 0 & 1 \end{bmatrix}$$

and a point

$$X = \begin{bmatrix} x \\ y \\ 1 \end{bmatrix}$$

the inverse of operation T is given by

$$T^{-1} = \begin{bmatrix} a & b & e \\ c & d & f \\ 0 & 0 & 1 \end{bmatrix}^{-1} = \begin{bmatrix} \left(\dfrac{d}{ad-bc} \right) & \left(\dfrac{-b}{ad-bc} \right) & \left(\dfrac{bf-de}{ad-bc} \right) \\ \left(\dfrac{-c}{ad-bc} \right) & \left(\dfrac{a}{ad-bc} \right) & \left(\dfrac{-af+ce}{ad-bc} \right) \\ 0 & 0 & 1 \end{bmatrix}$$

Multiplying, $TT^{-1} = T^{-1}T = I$. Details are left as an exercise.

Example
4.2.2

Given the linear transformation

$$T = \begin{bmatrix} 1 & 1 & 0 \\ 0 & 1 & 0 \\ 0 & 0 & 1 \end{bmatrix}$$

Using the formula in Theorem 4.2.3,

$$T^{-1} = \begin{bmatrix} 1 & -1 & 0 \\ 0 & 1 & 0 \\ 0 & 0 & 1 \end{bmatrix}$$

URL
4.2.7

Multiplying T by T^{-1} yields

$$\begin{bmatrix} 1 & -1 & 0 \\ 0 & 1 & 0 \\ 0 & 0 & 1 \end{bmatrix} \begin{bmatrix} 1 & -1 & 0 \\ 0 & 1 & 0 \\ 0 & 0 & 1 \end{bmatrix} = \begin{bmatrix} 1 & 0 & 0 \\ 0 & 1 & 0 \\ 0 & 0 & 1 \end{bmatrix}$$

The Image of a Line Under a Linear Transformation

Finding the image of a line under a given linear transformation involves additional considerations and produces an interesting result.

Theorem
4.2.4

The image of a line $u = [u_1 \quad u_2 \quad u_3]$ under a linear transformation

$$A = \begin{bmatrix} a & b & e \\ c & d & f \\ 0 & 0 & 1 \end{bmatrix}$$

is given by the equation $uA^{-1} = ku'$.

A line u may be represented in matrix form as

$$[u_1 \quad u_2 \quad u_3] \begin{bmatrix} x \\ y \\ 1 \end{bmatrix} = 0$$

The image of a line u under a linear transformation S may be represented as $uS = u'$. Since S is itself a linear transformation, it must have the same general form as transformation matrices for points

$$A = \begin{bmatrix} a & b & e \\ c & d & f \\ 0 & 0 & 1 \end{bmatrix}$$

The following argument derives S, the line transformation matrix, in terms of A, the point transformation matrix.

1. Since u and u' are lines, $uX = 0$ and $u'X' = 0$.
2. Then $uX = u'X'$.
3. Since X' is the image of X, $AX = X'$.

4. Substituting observation 3 in observation 2 yields $u(X) = u'(AX)$
   $\rightarrow (u)X = (u'A)X$.
5. Then $u = u'A$.
6. Multiplying both sides of the equation on the right by A^{-1} yields
   $uA^{-1} = ku'$, where $k \in \mathbb{R}$, the set of real numbers.
7. So, the line transformation matrix S is just the inverse of the point
   transformation matrix A.

**Example
4.2.3**

Given a linear transformation

$$T = \begin{bmatrix} 1 & 1 & 0 \\ 0 & 1 & 0 \\ 0 & 0 & 1 \end{bmatrix}$$

and a line $u = \begin{bmatrix} 1 & 2 & 1 \end{bmatrix}$, find the image of u under the linear transformation. According to Theorem 4.2.4, the image of u under this transformation may be computed as

$$\begin{bmatrix} 1 & 2 & 1 \end{bmatrix} \begin{bmatrix} 1 & -1 & 0 \\ 0 & 1 & 0 \\ 0 & 0 & 1 \end{bmatrix} \rightarrow k\begin{bmatrix} 1 & 1 & 1 \end{bmatrix}$$

In other words, the line with equation $x + 2y + 1 = 0$ is mapped onto the line with equation $x + y + 1 = 0$, or any multiple thereof. Table 4.2.1 summarizes the notations introduced in this section.

Summary

Because linear transformations are functions, students of mathematics will find many of their features familiar. For instance, linear transformations may be composed to create other linear transformations. Linear transformations have inverses, so they may be "undone." The concept of finding the image of a point under a given transformation corresponds directly with that of evaluating a function at a given point. On the other hand, the concept of finding an image of a line under a linear transformation is unfamiliar to most students.

Linear Transformation of a Point
Q: What is the image of the point $X = (x, y)$ under the linear transformation T given by

$$T = \begin{bmatrix} a & b & e \\ c & d & f \\ 0 & 0 & 1 \end{bmatrix}$$

A: $TX = X'$, or

$$\begin{bmatrix} a & b & e \\ c & d & f \\ 0 & 0 & 1 \end{bmatrix} \begin{bmatrix} x \\ y \\ 1 \end{bmatrix} = \begin{bmatrix} x' \\ y' \\ 1 \end{bmatrix}$$

Linear Transformation of a Line

Q: What is the image of the line $u = \begin{bmatrix} u_1 & u_2 & u_3 \end{bmatrix}$ under the linear transformation T given by

$$T = \begin{bmatrix} a & b & e \\ c & d & f \\ 0 & 0 & 1 \end{bmatrix}$$

A: $uT^{-1} = ku'$, or $\begin{bmatrix} u_1 & u_2 & u_3 \end{bmatrix} \begin{bmatrix} a & b & e \\ c & d & f \\ 0 & 0 & 1 \end{bmatrix}^{-1} = k\begin{bmatrix} u_1' & u_2' & u_3' \end{bmatrix}$,

where

$$\begin{bmatrix} a & b & e \\ c & d & f \\ 0 & 0 & 1 \end{bmatrix}^{-1} = \begin{bmatrix} \left(\dfrac{d}{ad-bc}\right) & \left(\dfrac{-b}{ad-bc}\right) & \left(\dfrac{bf-de}{ad-bc}\right) \\ \left(\dfrac{-c}{ad-bc}\right) & \left(\dfrac{a}{ad-bc}\right) & \left(\dfrac{-af+ce}{ad-bc}\right) \\ 0 & 0 & 1 \end{bmatrix}$$

URLs		Note: Begin each URL with the prefix http://
4.2.1	c	standards.nctm.org/
4.2.2	s	www.ScienceU.com/library/articles/isometries/translation.html
4.2.3	s	www.ScienceU.com/library/articles/isometries/rotation.html
4.2.4	s	www.ScienceU.com/library/articles/isometries/reflection.html
4.2.5	h	www-history.mcs.st-and.ac.uk/history/Mathematicians/Descartes.html
4.2.6	s	www.mkaz.com/math/matrix.html
4.2.7	s	www.mkaz.com/math/matrix.html

Table 4.2.1 Section 4.2 URLs (c = concept, h = history, s = software, d = data)

Exercises

1. Given transformation matrices

$$T = \begin{bmatrix} 1 & 0 & 2 \\ 0 & 1 & -3 \\ 0 & 0 & 1 \end{bmatrix} \quad \text{and} \quad S = \begin{bmatrix} 1 & 0 & 0 \\ 0 & -1 & 0 \\ 0 & 0 & 1 \end{bmatrix}$$

and a point and line

$$X = \begin{bmatrix} 3 \\ -1 \\ 1 \end{bmatrix} \qquad \text{and} \qquad u = \begin{bmatrix} 1 & -1 & 3 \end{bmatrix}, \qquad \text{find}$$

a) TX

b) SX

c) TS

d) ST

e) $T(SX)$

f) $(TS)X$

g) T^{-1}

h) TT^{-1}

i) S^{-1}

j) SS^{-1}

k) UT^{-1}

l) uS^{-1}

2. Sketch and label

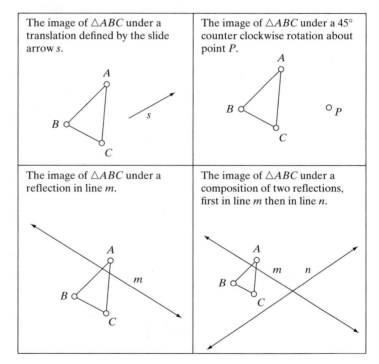

The image of $\triangle ABC$ under a translation defined by the slide arrow s.	The image of $\triangle ABC$ under a 45° counter clockwise rotation about point P.
The image of $\triangle ABC$ under a reflection in line m.	The image of $\triangle ABC$ under a composition of two reflections, first in line m then in line n.

3. Given a linear transformation T, the image of a point X under the transformation is given by the matrix equation $TX = X'$. The matrix T relates the coordinates of an object X to the coordinates of its image X' under the linear transformation. What matrix relates the parameters of an object line u to its image u' under the same linear transformation?

4. How many sets of object-image points are necessary to uniquely define a linear transformation? Why?

5. Prove Theorem 4.2.3.

6. Demonstrate by example that composition of linear transformations is not commutative.

7. Demonstrate by example that composition of linear transformations is associative.

Investigations

Translations (Slides)
Tool(s) Geometers Sketchpad   *Data File(s)* cd4_2_1.gsp

Focus
A translation (slide) moves an object a given distance in a given direction, often indicated by a slide arrow.

Tasks
1. Double click on the Translate *j* action key. Describe the action.
2. Measure the length and slope of *j* before and after the translation.
3. Measure *AA'* and *BB'*.
4. Sketch the segment *j*, the slide arrow, and an additional line *l* that would be invariant (slide along itself) under the given translation.
5. Double click on the Translate Triangle action key. Describe the action.
6. Measure the sides and angles of the triangle before and after the translation.
7. Compare the area of the triangle before and after the translation.
8. Measure *CC'* and *EE'*.
9. Sketch the triangle, the slide arrow, and an additional line *l* that would be invariant under the given translation.
10. What action would reverse a translation? Or, what is the inverse of a translation?

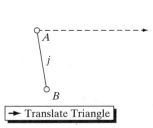

To undo any translation, just choose undo translation in the Edit menu.

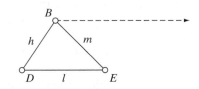

Reflections (Flips)
Tool(s) Geometers Sketchpad   *Data File(s)* cd4_2_2.gsp

Focus
A reflection produces a mirror image of an object on the opposite side of a reflecting line.

Tasks
1. Double click on Reflect Triangle. Double click again. What is the inverse of a reflection?
2. Measure and compare the distances from each point of the original triangle and reflected triangle to the reflecting line.
3. Sketch the original triangle, reflecting line, and at least one additional line that is invariant under this reflection.

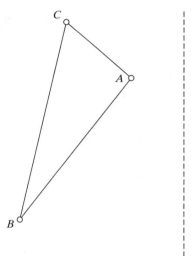

Reflecting Line

Rotations (Turns)
Tool(s) Geometers Sketchpad   *Data File(s)* **cd4_2_3.gsp**

Focus
A rotation is defined by a turn center and a turn angle.

Tasks
1. Double click on the Rotate Triangle action button. Double click again. What is the inverse of a rotation?
2. Describe a method for determining the turn angle of a rotation given an object and its image.
3. Identify any invariant points or lines associated with this rotation.

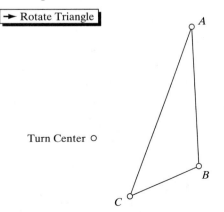

Glide Reflections
Tool(s) Geometers Sketchpad   *Data File(s)* cd4_2_5.gsp

Focus
Glide reflections involve both translation and reflection.

Tasks
1. Double click on the Reflect Triangle action key then the Glide Triangle action key. Double click on the Glide Triangle action key then the Reflect Triangle action key. Compare the final results obtained using both sequences of actions.
2. Identify any points or lines that are invariant under this glide reflection.

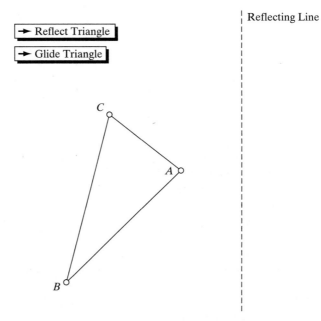

4.3   The Direct Isometries: Translations and Rotations

In this section, you will . . .

Use both geometric and matrix methods to
- **Represent translations and rotations;**
- **Identify invariant points and lines;**
- **Determine whether distance and area are preserved.**

Representing Translations

The matrix equation for the translation of a point is

$$\begin{bmatrix} 1 & 0 & e \\ 0 & 1 & f \\ 0 & 0 & 1 \end{bmatrix} \begin{bmatrix} x \\ y \\ 1 \end{bmatrix} = \begin{bmatrix} x + e \\ y + f \\ 1 \end{bmatrix} = \begin{bmatrix} x \\ y \\ 1 \end{bmatrix} + \begin{bmatrix} e \\ f \\ 1 \end{bmatrix} = \begin{bmatrix} x' \\ y' \\ 1 \end{bmatrix}$$

Figure 4.3.1 highlights two regions of the translation matrix. Each region provides different insights into the nature of the linear transformation under consideration and specific details of the motion produced. In this case, the 2×2 identity matrix in the upper left-hand corner of the matrix indicates that the motion does not in any way change the shape or orientation of objects in the plane. The elements in the shaded column are associated with shifts in the x- and y-coordinates of points in the plane. As seen in the image vector

$$\begin{bmatrix} x + e \\ y + f \\ 1 \end{bmatrix}$$

the x-coordinate of every point is shifted e units and the y-coordinate is shifted f units. This is exactly what one expects under the transformation of translation.

$$\begin{bmatrix} 1 & 0 & e \\ 0 & 1 & f \\ 0 & 0 & 1 \end{bmatrix} \begin{bmatrix} x \\ y \\ 1 \end{bmatrix} = \begin{bmatrix} x' \\ y' \\ 1 \end{bmatrix}$$

Figure 4.3.1 Translation Matrix

Conceptually, the inverse of a translation represented as

$$\begin{bmatrix} x + e \\ y + f \\ 1 \end{bmatrix}$$

should be a translation

$$\begin{bmatrix} x - e \\ y - f \\ 1 \end{bmatrix}$$

That is, the opposite of sliding an object a given amount in a specified direction should be a slide in the opposite direction by the same amount. This implies that

$$\begin{bmatrix} 1 & 0 & e \\ 0 & 1 & f \\ 0 & 0 & 1 \end{bmatrix}^{-1} = \begin{bmatrix} 1 & 0 & -e \\ 0 & 1 & -f \\ 0 & 0 & 1 \end{bmatrix}$$

Theorem 4.3.1   Given a translation

$$T = \begin{bmatrix} 1 & 0 & e \\ 0 & 1 & f \\ 0 & 0 & 1 \end{bmatrix}$$

the inverse transformation is given by

$$T^{-1} = \begin{bmatrix} 1 & 0 & -e \\ 0 & 1 & -f \\ 0 & 0 & 1 \end{bmatrix}$$

The proof of this theorem is left as an exercise.

Example 4.3.1   Given the translation matrix

$$T = \begin{bmatrix} 1 & 0 & 7 \\ 0 & 1 & -3 \\ 0 & 0 & 1 \end{bmatrix}$$

the inverse of the translation is given by

$$T^{-1} = \begin{bmatrix} 1 & 0 & -7 \\ 0 & 1 & 3 \\ 0 & 0 & 1 \end{bmatrix}$$

Multiplying these matrices yields the identity matrix

$$\begin{bmatrix} 1 & 0 & 7 \\ 0 & 1 & -3 \\ 0 & 0 & 1 \end{bmatrix} \begin{bmatrix} 1 & 0 & -7 \\ 0 & 1 & 3 \\ 0 & 0 & 1 \end{bmatrix} = \begin{bmatrix} 1 & 0 & 0 \\ 0 & 1 & 0 \\ 0 & 0 & 1 \end{bmatrix}$$

Invariance Under Translation

Experiences in the real world suggest that measurements of length, perimeter, angle, and area are preserved (invariant) under translations. Other features are not. For instance, when you slide an object on a table, you slide the entire object. If any point of the object were fixed (invariant), moving the rest of the object would require either distorting (stretching) or breaking the object. This geometric analysis leads to the conclusion that translations have no invariant points. An analytic approach based on matrix algebra leads to the same result.

**Theorem
4.3.2**

Nonidentity translations have no invariant points.

By definition, invariant points do not move under transformation. In other words,

$$\begin{bmatrix} x \\ y \\ 1 \end{bmatrix} = \begin{bmatrix} x' \\ y' \\ 1 \end{bmatrix}$$

In the case of translation, this implies that

$$\begin{bmatrix} 1 & 0 & e \\ 0 & 1 & f \\ 0 & 0 & 1 \end{bmatrix} \begin{bmatrix} x \\ y \\ 1 \end{bmatrix} = \begin{bmatrix} x \\ y \\ 1 \end{bmatrix}$$

Expanding and simplifying terms yields $x + e = x$ and $y + f = y$. This means that $e = 0$ and $f = 0$. Consequently, nonidentity translations have no invariant points.

Under some linear translations, lines are mapped onto themselves. In such circumstances, the lines are said to be invariant under the transformation. For instance, a transformation may slide points along a given line with a specified distance. While the points themselves are not invariant, the line they determine is invariant. In Figure 4.3.2, the image of $\triangle ABC$ under a translation defined by the indicated slide arrow is $\triangle A'B'C'$. The dotted lines formed by connecting each vertex with its image are invariant under the translation, as are all other lines parallel to the slide arrow.

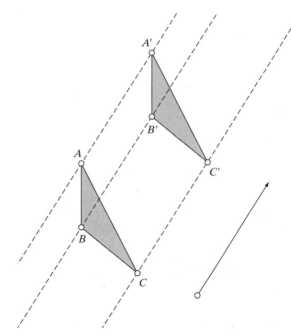

Figure 4.3.2 Invariant Lines

When a linear transformation is well understood, a geometric approach is generally the easiest and fastest approach for identifying invariant lines. When the transformation is not well understood, an analytic approach is needed for identifying invariant lines. The following theorem establishes a method for identifying invariant lines associated with any linear transformation.

Theorem 4.3.3

Every translation has invariant lines.

Invariant lines do not move under transformation, so $\lfloor u_1 \quad u_2 \quad u_3 \rfloor = k \lfloor u_1' \quad u_2' \quad u_3' \rfloor$, where $k \in R$ to allow for equivalent forms of the same line. Theorems 4.2.4 and 4.3.1 imply that

$$\begin{bmatrix} u_1 & u_2 & u_3 \end{bmatrix} \begin{bmatrix} 1 & 0 & -7 \\ 0 & 1 & 3 \\ 0 & 0 & 1 \end{bmatrix} = k \begin{bmatrix} u_1 & u_2 & u_3 \end{bmatrix}$$

Expanding and simplifying terms yields the following observations:

1. $u_1 = k u_1$
2. $u_2 = k u_2$
3. $u_1(-e) + u_2(-f) + u_3 = k u_3$

If $k = 1$, then observation 3 may be simplified to

$$u_2 = \left(\frac{-e}{f} \right) u_1$$

This leads to solutions of the form

$$\left[u_1 \quad \left(\frac{-e}{f} \right) u_1 \quad u_3 \right]$$

Rewriting this expression as

$$u_1 x + \left(\frac{-e u_1}{f} \right) y + u_3 = 0$$

and rearranging to slope-intercept form

$$y = \left(\frac{f}{e} \right) x + \left(\frac{f}{u_1 e} \right) u_3$$

it becomes clear that all invariant lines have the same slope as the slide arrow. The choice of u_3 is independent of u_1, leading to an infinite number of possible y-intercepts of the form

$$\left(\frac{f}{u_1 e} \right) u_3$$

So, the slide arrow and all lines parallel to it are invariant under the translation.

Example 4.3.2

Given the translation matrix

$$T = \begin{bmatrix} 1 & 0 & 7 \\ 0 & 1 & -3 \\ 0 & 0 & 1 \end{bmatrix}$$

find all invariant lines under the transformation. Using the approach of Theorem 4.3.4,

$$[u_1 \quad u_2 \quad u_3] \begin{bmatrix} 1 & 0 & -7 \\ 0 & 1 & 3 \\ 0 & 0 & 1 \end{bmatrix} = k[u_1 \quad u_2 \quad u_3]$$

Expanding and simplifying terms leads to

$$[u_1 \quad u_2 \quad (-7u_1 + 3u_2 + u_3)] = [ku_1 \quad ku_2 \quad ku_3].$$

Setting $k = 1$, we obtain the relationship

$$u_2 = \left(\frac{7}{3}\right)u_1$$

This leads to lines of the form

$$\left[u_1 \quad \left(\frac{7}{3}\right)u_1 \quad u_3\right]$$

Written in algebraic form and letting $u_1 = 1$, the equation of the line is

$$x + \left(\frac{7}{3}\right)y + u_3 \quad \text{or} \quad y = \left(\frac{-3}{7}\right)x + \left(\frac{-3}{7}\right)u_3$$

So all lines parallel to the slide arrow are invariant.

    A number of geometric features other than points and lines may be preserved under a given linear transformation, including distance between points and area of polygons. In the case of translations, this is expected.

Theorem 4.3.4

Distance is invariant under the transformation of translation.

The distance between two given points in the Euclidean plane (x_1, y_1) and (x_2, y_2) is given by

$$d = \sqrt{(x_2 - x_1)^2 + (y_2 - y_1)^2}$$

Computing the distance between corresponding image points (x_1', y_1') and (x_2', y_2') one obtains

$$d = \sqrt{([x_2 + e] - [x_1 + e])^2 + ([y_2 + f] - [y_1 + f])^2}$$

Simplifying this equation yields

$$d = \sqrt{(x_2 - x_1)^2 + (y_2 - y_1)^2}$$

So, distance is preserved under transformations.

Theorem 4.3.5

Area is invariant under the transformation of translation.

Referring to Theorem 4.1.1, the area determined by any three non-collinear points in the plane $(x_1, y_1), (x_2, y_2), (x_3, y_3)$ is

$$\text{Area} = \left(\frac{1}{2}\right) \text{abs}\left(\begin{vmatrix} x_1 & x_2 & x_3 \\ y_1 & y_2 & y_3 \\ 1 & 1 & 1 \end{vmatrix}\right)$$

After translation, the area determined by the images of these points is

$$\text{Area} = \left(\frac{1}{2}\right) \text{abs}\left(\begin{vmatrix} (x_1 + e) & (x_2 + e) & x_3 + e) \\ (y_1 + f) & (y_2 + f) & (y_3 + f) \\ 1 & 1 & 1 \end{vmatrix}\right)$$

Demonstrating that these two equations are equivalent is left as an exercise.

Example 4.3.3

Given the translation matrix

$$T = \begin{bmatrix} 1 & 0 & 7 \\ 0 & 1 & -3 \\ 0 & 0 & 1 \end{bmatrix}$$

and points $(1, 1), (2, 5)$, and $(6, 2)$, verify that the area determined by these points is preserved under translation T. By Equation 4.1.1, the area determined by these points is given by

$$\left(\frac{1}{2}\right) \text{abs}\left(\begin{vmatrix} 1 & 2 & 6 \\ 1 & 5 & 2 \\ 1 & 1 & 1 \end{vmatrix}\right) = 9.5$$

The image of the given points under T is given by

$$\begin{bmatrix} 1 & 0 & 7 \\ 0 & 1 & -3 \\ 0 & 0 & 1 \end{bmatrix}\begin{bmatrix} 1 & 2 & 6 \\ 1 & 5 & 2 \\ 1 & 1 & 1 \end{bmatrix} = \begin{bmatrix} 8 & 9 & 13 \\ -2 & 2 & -1 \\ 1 & 1 & 1 \end{bmatrix}$$

The triangular area determined by these points is given by

$$\left(\frac{1}{2}\right) \text{abs}\left(\begin{vmatrix} 8 & 9 & 13 \\ -2 & 2 & -1 \\ 1 & 1 & 1 \end{vmatrix}\right) = 9.5$$

So, the area determined by both triangles is the same.

Representing Rotations

The matrix equation for the rotation of a point through an angle θ about the origin is

$$\begin{bmatrix} a & b & 0 \\ c & d & 0 \\ 0 & 0 & 1 \end{bmatrix}\begin{bmatrix} x \\ y \\ 1 \end{bmatrix} = \begin{bmatrix} \cos\theta & -\sin\theta & 0 \\ \sin\theta & \cos\theta & 0 \\ 0 & 0 & 1 \end{bmatrix}\begin{bmatrix} x \\ y \\ 1 \end{bmatrix} = \begin{bmatrix} x' \\ y' \\ 1 \end{bmatrix}$$

where turn angles in the counterclockwise direction are positive. The 2×2 identity matrix in the upper left-hand corner of the matrix in Figure 4.3.3 reorients the object in the plane. The third column of the matrix indicates that the center of the rotation is the origin.

Thinking geometrically, the inverse of a rotation should be a rotation in the opposite direction about the same point, using the same turn angle in both cases.

$$\begin{bmatrix} a & b & 0 \\ c & d & 0 \\ 0 & 0 & 1 \end{bmatrix}\begin{bmatrix} x \\ y \\ 1 \end{bmatrix} = \begin{bmatrix} \cos\theta & -\sin\theta & 0 \\ \sin\theta & \cos\theta & 0 \\ 0 & 0 & 1 \end{bmatrix}\begin{bmatrix} x \\ y \\ 1 \end{bmatrix} = \begin{bmatrix} x' \\ y' \\ 1 \end{bmatrix}$$

Figure 4.3.3 Rotation Matrix

Theorem 4.3.6

Given a rotation about the origin

$$R = \begin{bmatrix} \cos\theta & -\sin\theta & 0 \\ \sin\theta & \cos\theta & 0 \\ 0 & 0 & 1 \end{bmatrix}$$

the inverse of R is given by

$$\begin{bmatrix} \cos\theta & -\sin\theta & 0 \\ \sin\theta & \cos\theta & 0 \\ 0 & 0 & 1 \end{bmatrix}^{-1} = \begin{bmatrix} \cos(-\theta) & -\sin(-\theta) & 0 \\ \sin(-\theta) & \cos(-\theta) & 0 \\ 0 & 0 & 1 \end{bmatrix}$$

$$= \begin{bmatrix} \cos\theta & \sin\theta & 0 \\ -\sin\theta & \cos\theta & 0 \\ 0 & 0 & 1 \end{bmatrix}$$

The proof of this theorem is left as an exercise.

Example 4.3.4

Find the image of the point $(1, 3)$ under a $30°$ counterclockwise rotation about the origin. The rotation matrix is given by

$$\begin{bmatrix} \cos\theta & -\sin\theta & 0 \\ \sin\theta & \cos\theta & 0 \\ 0 & 0 & 1 \end{bmatrix} = \begin{bmatrix} \dfrac{\sqrt{3}}{2} & \dfrac{-1}{2} & 0 \\ \dfrac{1}{2} & \dfrac{\sqrt{3}}{2} & 0 \\ 0 & 0 & 1 \end{bmatrix}$$

The image of the given point is given by

$$
\begin{bmatrix} \dfrac{\sqrt{3}}{2} & \dfrac{-1}{2} & 0 \\ \dfrac{1}{2} & \dfrac{\sqrt{3}}{2} & 0 \\ 0 & 0 & 1 \end{bmatrix}
\begin{bmatrix} 1 \\ 3 \\ 1 \end{bmatrix} =
\begin{bmatrix} \dfrac{\sqrt{3}-3}{2} \\ \dfrac{3\sqrt{3}+1}{2} \\ 1 \end{bmatrix}
$$

The inverse operation may be applied to the image point to return to the original location. In matrix form, this operation is represented as

$$
\begin{bmatrix} \dfrac{\sqrt{3}}{2} & \dfrac{1}{2} & 0 \\ \dfrac{-1}{2} & \dfrac{\sqrt{3}}{2} & 0 \\ 0 & 0 & 1 \end{bmatrix}
\begin{bmatrix} \dfrac{\sqrt{3}-3}{2} \\ \dfrac{3\sqrt{3}+1}{2} \\ 1 \end{bmatrix} =
\begin{bmatrix} 1 \\ 3 \\ 1 \end{bmatrix}
$$

Invariance Under Rotation

Thinking geometrically, there is at least one point that should be invariant under a rotation, the center of the rotation. Since every other point in the plane undergoes a change of position determined by its distance from the turn center and the turn angle, no other points are invariant. This fact is easily demonstrated using an analytic approach.

Theorem 4.3.7

The origin is invariant under the rotation

$$
R = \begin{bmatrix} \cos\theta & -\sin\theta & 0 \\ \sin\theta & \cos\theta & 0 \\ 0 & 0 & 1 \end{bmatrix}
$$

The image of the origin under the rotations is given by

$$
\begin{bmatrix} \cos\theta & -\sin\theta & 0 \\ \sin\theta & \cos\theta & 0 \\ 0 & 0 & 1 \end{bmatrix}
\begin{bmatrix} 0 \\ 0 \\ 1 \end{bmatrix} =
\begin{bmatrix} 0 \\ 0 \\ 1 \end{bmatrix}
$$

To show that the origin is the only invariant point (when θ is not a multiple of 2π), expand the matrix equation

$$
\begin{bmatrix} \cos\theta & -\sin\theta & 0 \\ \sin\theta & \cos\theta & 0 \\ 0 & 0 & 1 \end{bmatrix}
\begin{bmatrix} x \\ y \\ 1 \end{bmatrix} =
\begin{bmatrix} x \\ y \\ 1 \end{bmatrix}
$$

to obtain $x\cos\theta - y\sin\theta = x$ and $x\sin\theta + y\cos\theta = y$. Solving both equations for y yields

$$
y = \frac{x(\cos\theta - 1)}{\sin\theta} = \frac{-x\sin\theta}{\cos\theta - 1}
$$

Setting the expressions equal to one another and cross multiplying produces the equation $x(\cos \theta - 1)^2 = -x(\sin \theta)^2$. Since $(\cos \theta - 1)^2$ and $(\sin \theta)^2$ are both positive, the only value for x satisfying the equation is $x = 0$. Substituting this value into either expression for y forces $y = 0$. So, the only invariant point is the origin.

Thinking geometrically, every line in the plane is reoriented under a rotation, so there should be no invariant lines other than multiples of 180°. The following analysis confirms that impression.

Theorem 4.3.8   Rotations (where θ is not a multiple of π) do not have invariant lines.

Invariant lines do not move under transformation, so $[u_1 \;\; u_2 \;\; u_3] = k[u_1' \;\; u_2' \;\; u_3']$, where $k \in R$ to allow for equivalent forms of the same line. Theorem 4.3.1 implies that

$$[u_1 \;\; u_2 \;\; u_3] \begin{bmatrix} \cos \theta & \sin \theta & 0 \\ -\sin \theta & \cos \theta & 0 \\ 0 & 0 & 1 \end{bmatrix} = k[u_1 \;\; u_2 \;\; u_3]$$

Expanding and simplifying terms yields the following observations:

1. $u_1 \cos \theta - u_2 \sin \theta = ku_1$
2. $u_1 \sin \theta + u_2 \cos \theta = ku_2$
3. $u_3 = ku_3$

If $u_3 \neq 0$, $k = 1$. Then observation 1 may be simplified to

$$u_2 = \left(\frac{\cos \theta - 1}{\sin \theta} \right) u_1, \text{ where } \theta \neq n\pi$$

This leads to solutions of the form

$$\left[u_1 \;\; \left(\frac{\cos \theta - 1}{\sin \theta} \right) u_1 \;\; u_3 \right]$$

The coefficient of the second term is clearly not a constant, so the object described cannot be a straight line. No lines are preserved under the transformation of rotation.

Rotations are common transformations in everyday life. Our experience suggests that distance and area are preserved under rotations. The following theorems formalize those impressions.

Theorem 4.3.9   Distance is invariant under the transformation of rotation.

A proof of this theorem is left as an exercise.

Example 4.3.5   Find the distance between the points $(1, 0)$ and $(0, 1)$ and their images under a 90° counterclockwise rotation. Using the distance formula, the distance between $(1, 0)$ and $(0, 1)$ may be computed to be $\sqrt{2}$. The image of the points under the rotation is given by the expression

$$\begin{bmatrix} 0 & -1 & 0 \\ 1 & 0 & 0 \\ 0 & 0 & 1 \end{bmatrix} \begin{bmatrix} 1 & 0 \\ 0 & 1 \\ 1 & 1 \end{bmatrix} = \begin{bmatrix} 0 & -1 \\ 1 & 0 \\ 1 & 1 \end{bmatrix}$$

Using the distance formula on these points also yields $\sqrt{2}$. So, the distance between points, before and after rotation, is the same.

Theorem 4.3.10 Area is invariant under the transformation of rotation.

A proof of this theorem is left as an exercise.

Example 4.3.6 Given the rotation matrix

$$R = \begin{bmatrix} 0 & -1 & 0 \\ 1 & 0 & 0 \\ 0 & 0 & 1 \end{bmatrix}$$

and points $(1, 1)$, $(2, 5)$, and $(6, 2)$, verify that the area determined by these points is preserved under rotation R. By Equation 4.1.1, the area determined by these points is given by

$$\left(\frac{1}{2}\right)\text{abs}\left(\begin{vmatrix} 1 & 2 & 6 \\ 1 & 5 & 2 \\ 1 & 1 & 1 \end{vmatrix}\right) = 9.5$$

The image of the given points under R is given by

$$\begin{bmatrix} 0 & -1 & 0 \\ 1 & 0 & 0 \\ 0 & 0 & 1 \end{bmatrix} \begin{bmatrix} 1 & 2 & 6 \\ 1 & 5 & 2 \\ 1 & 1 & 1 \end{bmatrix} = \begin{bmatrix} -1 & -5 & -2 \\ 1 & 2 & 6 \\ 1 & 1 & 1 \end{bmatrix}$$

The triangular area determined by these points is given by

$$\left(\frac{1}{2}\right)\text{abs}\left(\begin{vmatrix} -1 & -5 & -2 \\ 1 & 2 & 6 \\ 1 & 1 & 1 \end{vmatrix}\right) = 9.5$$

So, the area determined by both triangles is the same.

Summary

Translations and rotations together make up what are known as the direct isometries. Direct isometries preserve distance and area. As a result, the images and objects differ only in their position or orientation in the Euclidean plane. Nonidentity translations have no invariant points and infinitely many invariant lines. Rotations (where θ is not a multiple of π) have one invariant point and no invariant lines.

Translation of a Point

Q: What is the image of the point $X(x, y, 1)$ under the translation given by

$$T = \begin{bmatrix} 1 & 0 & e \\ 0 & 1 & f \\ 0 & 0 & 1 \end{bmatrix}$$

A: A point $X'(x', y', 1)$

$$\begin{bmatrix} 1 & 0 & e \\ 0 & 1 & f \\ 0 & 0 & 1 \end{bmatrix} \begin{bmatrix} x \\ y \\ 1 \end{bmatrix} = \begin{bmatrix} x + e \\ y + f \\ 1 \end{bmatrix} = \begin{bmatrix} x' \\ y' \\ 1 \end{bmatrix}$$

Inverse of a Translation Matrix

Q: What is the inverse of a given translation matrix?

A: A translation represented by the inverse of the given translation's matrix.

$$\begin{bmatrix} 1 & 0 & e \\ 0 & 1 & f \\ 0 & 0 & 1 \end{bmatrix}^{-1} = \begin{bmatrix} 1 & 0 & -e \\ 0 & 1 & -f \\ 0 & 0 & 1 \end{bmatrix}$$

Translation of a Line

Q: What is the image of the line $\begin{bmatrix} u_1 & u_2 & u_3 \end{bmatrix}$ under the translation given by

$$T = \begin{bmatrix} 1 & 0 & e \\ 0 & 1 & f \\ 0 & 0 & 1 \end{bmatrix}$$

A: A line $\begin{bmatrix} u'_1 & u'_2 & u'_3 \end{bmatrix}$

$$\begin{bmatrix} u_1 & u_2 & u_3 \end{bmatrix} \begin{bmatrix} 1 & 0 & -e \\ 0 & 1 & -f \\ 0 & 0 & 1 \end{bmatrix} = k\begin{bmatrix} u'_1 & u'_2 & u'_3 \end{bmatrix}$$

Invariant Lines Under a Translation

Q: How are invariant lines under a translation determined?

A: Using the line equation, set the image line equal to the object line.

$$\begin{bmatrix} u_1 & u_2 & u_3 \end{bmatrix} \begin{bmatrix} 1 & 0 & -e \\ 0 & 1 & -f \\ 0 & 0 & 1 \end{bmatrix} = k\begin{bmatrix} u_1 & u_2 & u_3 \end{bmatrix}$$

Rotation of a Point about the Origin

Q: What is the image of the point $X(x, y, 1)$ under a rotation of $\theta°$ about the origin?

A: A point $X'(x', y', 1)$

$$\begin{bmatrix} a & b & 0 \\ c & d & 0 \\ 0 & 0 & 1 \end{bmatrix}\begin{bmatrix} x \\ y \\ 1 \end{bmatrix} = \begin{bmatrix} \cos\theta & -\sin\theta & 0 \\ \sin\theta & \cos\theta & 0 \\ 0 & 0 & 1 \end{bmatrix}\begin{bmatrix} x \\ y \\ 1 \end{bmatrix} = \begin{bmatrix} x' \\ y' \\ 1 \end{bmatrix}$$

Inverse of a Rotation Matrix

Q: What is the inverse of a rotation?

A: A rotation represented by the inverse of the rotation matrix.

$$\begin{bmatrix} \cos\theta & -\sin\theta & 0 \\ \sin\theta & \cos\theta & 0 \\ 0 & 0 & 1 \end{bmatrix}^{-1} = \begin{bmatrix} \cos\theta & \sin\theta & 0 \\ -\sin\theta & \cos\theta & 0 \\ 0 & 0 & 1 \end{bmatrix}$$

Rotation of a Line about the Origin

Q: What is the image of the line $\begin{bmatrix} u_1 & u_2 & u_3 \end{bmatrix}$ under the rotation R given by

$$R = \begin{bmatrix} \cos\theta & -\sin\theta & 0 \\ \sin\theta & \cos\theta & 0 \\ 0 & 0 & 1 \end{bmatrix}$$

A: A line $\begin{bmatrix} u_1' & u_2' & u_3' \end{bmatrix}$

$$\begin{bmatrix} u_1 & u_2 & u_3 \end{bmatrix}\begin{bmatrix} \cos\theta & \sin\theta & 0 \\ -\sin\theta & \cos\theta & 0 \\ 0 & 0 & 1 \end{bmatrix} = k\begin{bmatrix} u_1' & u_2' & u_3' \end{bmatrix}$$

Exercises

1. Using geometric thinking, sketch a translation based on the matrix

$$T = \begin{bmatrix} 1 & 0 & 4 \\ 0 & 1 & -1 \\ 0 & 0 & 1 \end{bmatrix}$$

2. Using geometric thinking, sketch a rotation based on the matrix

$$R = \begin{bmatrix} 0 & -1 & 0 \\ 1 & 0 & 0 \\ 0 & 0 & 1 \end{bmatrix}$$

3. Using matrix methods, find the image of the point $X(1, 5, 1)$ under a 45° rotation about the origin.

4. Using both geometric thinking and matrix methods, find all invariant points and lines under the translation given by the matrix

$$T = \begin{bmatrix} 1 & 0 & 4 \\ 0 & 1 & -1 \\ 0 & 0 & 1 \end{bmatrix}$$

5. Using both geometric thinking and matrix methods, find all invariant points and lines under the rotation given by the matrix

$$R = \begin{bmatrix} 0 & -1 & 0 \\ 1 & 0 & 0 \\ 0 & 0 & 1 \end{bmatrix}$$

6. Use both geometric thinking and matrix methods to explain the existence and nature of inverse translations and rotations. How does each approach complement the other?

7. Prove Theorem 4.3.1.

8. Prove Theorem 4.3.6.

9. Prove Theorem 4.3.9.

10. Prove Theorem 4.3.10.

11. Prove that angular measure is preserved under translations and rotations.

4.4 Indirect Isometries: Reflections

In this section, you will . . .

Use both geometric and matrix methods to:
- **Represent reflections;**
- **Identify invariant points and lines;**
- **Determine whether distance and area are preserved.**

Representing Reflections

$$\begin{bmatrix} 1 & 0 & 0 \\ 0 & -1 & 0 \\ 0 & 0 & 1 \end{bmatrix} \begin{bmatrix} x \\ y \\ 1 \end{bmatrix} = \begin{bmatrix} x' \\ y' \\ 1 \end{bmatrix}$$

Figure 4.4.1 Reflection Matrix

The matrix equation for the reflection (see Figure 4.4.1) of a point in the x-axis is given by

$$\begin{bmatrix} 1 & 0 & 0 \\ 0 & -1 & 0 \\ 0 & 0 & 1 \end{bmatrix} \begin{bmatrix} x \\ y \\ 1 \end{bmatrix} = \begin{bmatrix} x \\ -y \\ 1 \end{bmatrix} = \begin{bmatrix} x' \\ y' \\ 1 \end{bmatrix}$$

Under this transformation, every point is mapped onto the point having the same x-coordinate but the opposite y-coordinate. As a result, every point above the x-axis is mapped to a corresponding point below the x-axis, and vice versa.

The matrix equation of a reflection in the y-axis is

$$\begin{bmatrix} -1 & 0 & 0 \\ 0 & 1 & 0 \\ 0 & 0 & 1 \end{bmatrix} \begin{bmatrix} x \\ y \\ 1 \end{bmatrix} = \begin{bmatrix} -x \\ y \\ 1 \end{bmatrix} = \begin{bmatrix} x' \\ y' \\ 1 \end{bmatrix}$$

Notice that under this transformation, it is the x-coordinates that change sign.

Example 4.4.1

Find the image of the point $(1, 3)$ under a reflection in the x-axis. The matrix equation is given by

$$\begin{bmatrix} 1 & 0 & 0 \\ 0 & -1 & 0 \\ 0 & 0 & 1 \end{bmatrix} \begin{bmatrix} 1 \\ 3 \\ 1 \end{bmatrix} = \begin{bmatrix} 1 \\ -3 \\ 1 \end{bmatrix}$$

Thinking geometrically, an object and its image under a reflection will be congruent. There is reversal in orientation, however (see Figure 4.4.2). Starting at A and moving around the object in a clockwise direction, the vertices are, in order, A-B-C. Starting at A' and repeating this procedure, the vertices are, in order, A'-C'-B'. No combination of translations and rotations can produce this result.

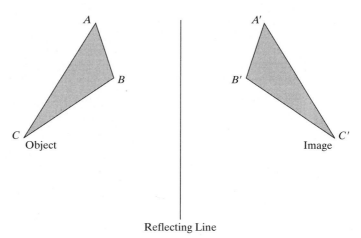

Figure 4.4.2 Reversal of Orientation

Draw a second object triangle intersected by the reflecting line. Find the image of that triangle under the transformation.

This distinctive difference between the transformation of reflection and the transformations of translation and rotation is highly significant. For instance, both translations and rotations may be accomplished using a composition of reflections. Figure 4.4.3 shows two parallel reflecting lines, *j* and *k*, and a series of three triangles. Considering Triangle 1 to be the object, its reflection in line *j* is Triangle 2. The reflection of Triangle 2 in line *k* is Triangle 3. Triangle 3 is clearly a translation of Triangle 1. Using similar logic, Triangle 1 may be viewed as the image of Triangle 3 under two successive reflections, first in line *k* then in line *j*. There is a simple relationship between the length of the slide arrow and the distance between the reflecting lines. Finding that relationship is left as an exercise.

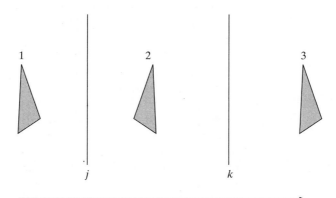

Figure 4.4.3 Translation as a Composition of Reflections

Draw a second object triangle intersected by the reflecting line *j*. Find the image of that triangle under the composition of transformations.

Figure 4.4.4 shows two intersecting reflecting lines, *j* and *k*, and a series of three triangles. Considering Triangle 1 to be the object, its reflection in line *j* is Triangle 2. The reflection of Triangle 2 in line *k* is Triangle 3. Triangle 3 is clearly a rotation of Triangle 1. Using similar logic, Triangle 1 may be viewed as the image of Triangle 3 under two successive reflections, first in line *k* then in line *j*. As in the case of parallel reflecting lines, there is a simple relationship between the turn angle of the rotation and the angle formed by the two reflecting lines. Finding the relationship is left as an exercise.

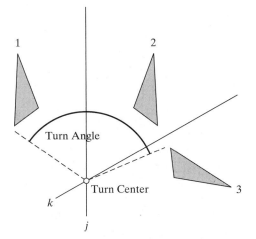

Figure 4.4.4 Rotation as a Composition of Reflections

Draw a second object triangle intersected by the reflecting line *j*. Find the image of that triangle under the composition of transformations.

The fact that both translations and rotations may be written as a composition of reflections, while no combination of translations and/or reflections is equivalent to a reflection, establishes reflections as a sort of "atom" of isometry. All isometries may be written as a composition of reflections. For this reason, reflections are mathematically more fundamental than translations or rotations.

Reflections are also more interesting and entertaining in a personal sense. For instance, Figure 4.4.5 shows three images of the author. The Normal Photo is of the author facing slightly off center to the camera. Two "Left" Faces was created by reflecting the left side of the Normal Photo over a center line. Two "Right" Faces was created by reflecting the right side over the same center line. The amusement that we experience when we encounter playful distortions of this sort is common enough that carnivals routinely include "Fun Houses" full of irregular mirrors. Playing with images in this way is a challenging and enjoyable way to develop mathematical insights into the nature of linear transformations.

**CD
4.4.1**

Normal Photo   Two "Left" Faces   Two "Right" Faces

Figure 4.4.5 Reflections and Faces

The CD-ROM Activity cd4_4_1 provides directions for creating playing with faces in this manner.

Thinking geometrically, it is clear that every reflection is its own inverse. The transformation of reflecting a given point in a given line is "undone" by reflecting it back again. The following theorem provides a notation.

**Theorem
4.4.1**

Given a reflection in the x-axis

$$F = \begin{bmatrix} 1 & 0 & 0 \\ 0 & -1 & 0 \\ 0 & 0 & 1 \end{bmatrix}$$

the inverse of F is given by

$$\begin{bmatrix} 1 & 0 & 0 \\ 0 & -1 & 0 \\ 0 & 0 & 1 \end{bmatrix}^{-1} = \begin{bmatrix} 1 & 0 & 0 \\ 0 & -1 & 0 \\ 0 & 0 & 1 \end{bmatrix}$$

The proof of this theorem is left as an exercise.

Invariance Under Reflection

The existence of invariant points and lines under the transformation of reflection may be investigated both geometrically and analytically. Thinking geometrically, it is clear that points on the reflecting line do not move under the transformation. They are invariant. Because every point on the reflecting line is *fixed* under this transformation, the line is said to be *point-wise* invariant. In addition, all lines perpendicular to the reflecting line are also invariant, though not point-wise invariant. For instance, in the case of a reflection in the x-axis, imagine all of the points on the line $x = 1$ sliding over one another as they cross the x-axis to assume their new locations, all the while remaining collinear. So, while the individual points are not invariant, the line on which they lie undergoes no motion. It is invariant. It is significant that the two sets of invariant lines are *orthogonal,* or perpendicular.

Using analytic techniques, the existence of invariant points and lines associated with a transformation matrix may be investigated without knowing the geometric features of the transformation.

Theorem 4.4.2

Points on the x-axis are invariant under the reflection

$$R = \begin{bmatrix} 1 & 0 & 0 \\ 0 & -1 & 0 \\ 0 & 0 & 1 \end{bmatrix}$$

For all invariant points $(x, y, 1)$

$$\begin{bmatrix} 1 & 0 & 0 \\ 0 & -1 & 0 \\ 0 & 0 & 1 \end{bmatrix} \begin{bmatrix} x \\ y \\ 1 \end{bmatrix} = \begin{bmatrix} x \\ y \\ 1 \end{bmatrix}$$

Expanding and simplifying this equation yields $x = x$ and $-y = y$. The interpretation of this finding is that x is unrestricted but $y = 0$. In other words, all points of the form $(x, 0, 1)$ are invariant. All such points lie on the x-axis.

Theorem 4.4.3

Both the x-axis and all lines perpendicular to it are invariant under the reflection

$$R = \begin{bmatrix} 1 & 0 & 0 \\ 0 & -1 & 0 \\ 0 & 0 & 1 \end{bmatrix}$$

Invariant lines do not move under transformation, so

$$\begin{bmatrix} u_1 & u_2 & u_3 \end{bmatrix} = k\begin{bmatrix} u_1' & u_2' & u_3' \end{bmatrix}$$

where $k \in R$ to allow for equivalent forms of the same line. Then

$$\begin{bmatrix} u_1 & u_2 & u_3 \end{bmatrix} \begin{bmatrix} 1 & 0 & 0 \\ 0 & -1 & 0 \\ 0 & 0 & 1 \end{bmatrix} = k\begin{bmatrix} u_1 & u_2 & u_3 \end{bmatrix}$$

Expanding and simplifying terms yields the following observations:

1. $u_1 = ku_1$
2. $-u_2 = ku_2$
3. $u_3 = ku_3$

If $k = 1$, then observation 2 may be simplified to $u_2 = 0$. This leads to solutions of the form

$$\begin{bmatrix} u_1 & 0 & u_3 \end{bmatrix} \begin{bmatrix} x \\ y \\ 1 \end{bmatrix} = 0$$

This may also be written as $x = -u_3/u_1$. This is the equation of a line perpendicular to the x-axis. Because u_3 is unrestricted, the equation represents all lines perpendicular to the x-axis. If $k = -1$, a similar analysis leads to solutions of the form

$$\begin{bmatrix} 0 & u_3 & 0 \end{bmatrix} \begin{bmatrix} x \\ y \\ 1 \end{bmatrix} = 0$$

which may also be written as $y = 0$. This is the equation of the x-axis.

Theorem 4.4.4

Distance is invariant under the transformation of reflection.

A proof of this theorem is left as an exercise.

Theorem 4.4.5

Area is invariant under the transformation of reflection.

A proof of this theorem is left as an exercise.

Summary

Reflections in the x-axis have a variety of invariant features. The axis of reflection is point-wise invariant, meaning every point on the axis is fixed. As a result, the line containing the axis is also invariant. All lines perpendicular to this axis are invariant as well, though not point-wise. Reflection also preserves distance and area. Angular measures before and after transformation have the same magnitude but the opposite sign. That is, clockwise angles become counterclockwise and vice versa. This effect produces a reversal in orientation unattainable by translation or rotation.

Reflection of a Point

Q: What is the image of the point $X(x, y, 1)$ under a reflection in the x-axis given by

$$T = \begin{bmatrix} 1 & 0 & 0 \\ 0 & -1 & 0 \\ 0 & 0 & 1 \end{bmatrix}$$

A: A point $X'(x', y', 1)$

$$\begin{bmatrix} 1 & 0 & 0 \\ 0 & -1 & 0 \\ 0 & 0 & 1 \end{bmatrix} \begin{bmatrix} x \\ y \\ 1 \end{bmatrix} = \begin{bmatrix} x \\ -y \\ 1 \end{bmatrix}$$

Inverse of a Reflection Matrix

Q: What is the inverse of a reflection in the x-axis?
A: A reflection in the x-axis.

$$\begin{bmatrix} 1 & 0 & 0 \\ 0 & -1 & 0 \\ 0 & 0 & 1 \end{bmatrix}^{-1} = \begin{bmatrix} 1 & 0 & 0 \\ 0 & -1 & 0 \\ 0 & 0 & 1 \end{bmatrix}$$

Reflection of a Line

Q: What is the image of the line $[u_1 \quad u_2 \quad u_3]$ under a reflection in the x-axis given by

$$T = \begin{bmatrix} 1 & 0 & 0 \\ 0 & -1 & 0 \\ 0 & 0 & 1 \end{bmatrix}$$

A: A line $[u_1' \quad u_2' \quad u_3']$

$$[u_1 \quad u_2 \quad u_3] \begin{bmatrix} 1 & 0 & 0 \\ 0 & -1 & 0 \\ 0 & 0 & 1 \end{bmatrix} = k[u_1' \quad u_2' \quad u_3']$$

Invariant Points Under a Reflection

Q: How are invariant points under a reflection in the x-axis determined?

A: Using the point equation, set the image point equal to the object point.

$$\begin{bmatrix} 1 & 0 & 0 \\ 0 & -1 & 0 \\ 0 & 0 & 1 \end{bmatrix} \begin{bmatrix} x \\ y \\ 1 \end{bmatrix} = \begin{bmatrix} x \\ y \\ 1 \end{bmatrix}$$

Invariant Lines Under a Reflection

Q: How are invariant lines under a reflection in the x-axis determined?

A: Using the line equation, set the image line equal to the object line.

$$[u_1 \quad u_2 \quad u_3] \begin{bmatrix} 1 & 0 & 0 \\ 0 & -1 & 0 \\ 0 & 0 & 1 \end{bmatrix} = k[u_1 \quad u_2 \quad u_3]$$

Exercises

1. Using geometric thinking, sketch a reflection of the triangle with vertices $(1, 1)$, $(4, 2)$, and $(2, 5)$ in the y-axis. Write the transformation matrix for a reflection in the y-axis.

2. Using geometric thinking, sketch a reflection of the triangle with vertices $(1, 1)$, $(4, 2)$, and $(2, 5)$ in the line $y = x$. Write the transformation matrix for a reflection in the line $y = x$.

3. Using matrix methods, find the image of the point $X(1, 5, 1)$ under a reflection in the y-axis.

4. Using matrix methods, find the image of the point $X(1, 5, 1)$ under a reflection in the line $y = x$.

5. Using both geometric thinking and matrix methods, find all invariant points and lines under a reflection in the y-axis.

6. Using both geometric thinking and matrix methods, find all invariant points and lines under a reflection in the line $y = x$.

7. Use both geometric thinking and matrix methods to explain the existence and nature of inverse reflections. How does each approach complement the other?

8. Given a translation T, a rotation R, and a reflection F, explain how the three transformations could be composed into a single linear transformation $S = T(R(F(X)))$. What relationship do you expect between the invariant points and lines of each transformation and the invariant points and lines of the composition, S?

9. Prove Theorem 4.4.1.

10. Prove Theorem 4.4.4.

11. Prove Theorem 4.4.5.

12. Demonstrate by example the relationship between the reflecting lines in Figure 4.4.3.

13. Demonstrate by example the relationship between the reflecting lines in Figure 4.4.4.

4.5 Composition and Analysis of Transformations

In this section,
you will . . .

- **Represent rotations about points other than the origin using a composition of transformations;**
- **Represent reflections in lines other than the *x*-axis using a composition of transformations;**
- **Determine the matrix of a linear transformation given corresponding points in its object and image sets;**
- **Relate what you have learned so far about invariant lines to the eigen equation, eigenvalues, and eigenvectors.**

Representing Rotations with Compositions of Matrices

Most of the linear transformations considered to this point are special cases: rotations about the origin, as opposed to any point *P*; reflections in the *x*-axis, instead of any line *r*; and so on. The matrices associated with these special cases are simple compared to most. In this section, we examine the use of simple matrices as building blocks in the construction of more complex transformations and their matrices. The general strategy employed is to express the desired transformation as a composition of simple matrices and their inverses. For instance, a rotation of the plane about the point (e, f) may be achieved through the composition of three Euclidean motions:

T: Translate the turn center (e, f) to the origin.
R: Rotate about the turn center, now at the origin.
T^{-1}: Translate the turn center back to its original location, (e, f).

For a given point set X, the composition may be written $T^{-1}RTX = X'$, or

$$\begin{bmatrix} 1 & 0 & e \\ 0 & 1 & f \\ 0 & 0 & 1 \end{bmatrix} \begin{bmatrix} \cos\theta & -\sin\theta & 0 \\ \sin\theta & \cos\theta & 0 \\ 0 & 0 & 1 \end{bmatrix} \begin{bmatrix} 1 & 0 & -e \\ 0 & 1 & -f \\ 0 & 0 & 1 \end{bmatrix} \begin{bmatrix} x \\ y \\ 1 \end{bmatrix} = \begin{bmatrix} x' \\ y' \\ 1 \end{bmatrix}$$

Multiplying the three transformation matrices together from left to right yields

$$\begin{bmatrix} \cos\theta & -\sin\theta & e(1 - \cos\theta) + f\sin\theta \\ \sin\theta & \cos\theta & -e\sin\theta + f(1 - \cos\theta) \\ 0 & 0 & 1 \end{bmatrix} \begin{bmatrix} x \\ y \\ 1 \end{bmatrix} = \begin{bmatrix} x' \\ y' \\ 1 \end{bmatrix}$$

This is the general formula for rotation about a point (e, f) and is formalized in Theorem 4.5.1. Given the complexity of the third column of the matrix, many students of mathematics prefer approaching generalized rotations through a composition of matrices.

Theorem
4.5.1

Given turn center $(e, f, 1)$ and turn angle θ, a rotation transformation is given by

$$\begin{bmatrix} \cos\theta & -\sin\theta & e(1-\cos\theta)+f\sin\theta \\ \sin\theta & \cos\theta & -e\sin\theta+f(1-\cos\theta) \\ 0 & 0 & 1 \end{bmatrix}\begin{bmatrix} x \\ y \\ 1 \end{bmatrix} = \begin{bmatrix} x' \\ y' \\ 1 \end{bmatrix}$$

A proof of this theorem is left as an exercise.

Representing Reflections with Compositions of Matrices

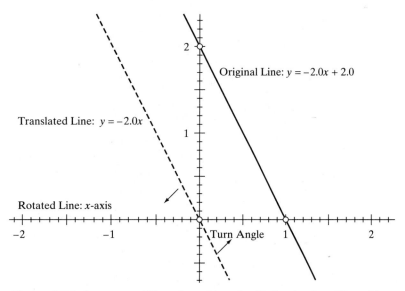

Figure 4.5.1 Sequence of Transformations for Reflecting in a Given Line

A similar approach may be taken to represent a reflection in a line other than the x-axis. Figure 4.5.1 shows such a line with equation $y = -2.0x + 2.0$. First, the given line is translated T so that it intersects the origin. It is then rotated R in a counterclockwise direction to coincide with the x-axis. After reflecting F in the x-axis, a rotation R^{-1} and translation T^{-1} return the reflecting line to its original position. The composition of these transformations may be written as $T^{-1}R^{-1}FRTX = X'$. The associated matrix equation is

$$\begin{bmatrix} 1 & 0 & 0 \\ 0 & 1 & 2 \\ 0 & 0 & 1 \end{bmatrix}\begin{bmatrix} \cos\theta & \sin\theta & 0 \\ -\sin\theta & \cos\theta & 0 \\ 0 & 0 & 1 \end{bmatrix}\begin{bmatrix} 1 & 0 & 0 \\ 0 & -1 & 0 \\ 0 & 0 & 1 \end{bmatrix}\begin{bmatrix} \cos\theta & -\sin\theta & 0 \\ \sin\theta & \cos\theta & 0 \\ 0 & 0 & 1 \end{bmatrix}\begin{bmatrix} 1 & 0 & 0 \\ 0 & 1 & -2 \\ 0 & 0 & 1 \end{bmatrix}\begin{bmatrix} x \\ y \\ 1 \end{bmatrix} = \begin{bmatrix} x' \\ y' \\ 1 \end{bmatrix}$$

where the turn angle $\theta = \arctan(2)$.

Using tools such as the TI-92 calculator or computer software packages such as Maple and Matlab, formulations of this sort are relatively easy to simplify and evaluate. The power of this procedure lies in its generality. Working with a small "tool kit" of basic transformations, one may reliably construct matrix representations for more complex transformations and justify the composition easily to others. Using this approach, a transformation matrix was determined for a reflection in the line $l\,[a \quad b \quad c]$, where l intersects the y-axis in the point $-c/b$ with slope $-a/b$, $b \neq 0$.

Theorem
4.5.2

Given a line $l\,[a \quad b \quad c]$, the transformation matrix R_1 for a reflection in l is given by

CD
4.5.1

$$R_1 = \begin{bmatrix} \dfrac{-(a^2 - b^2)}{a^2 + b^2} & \dfrac{-2ab}{a^2 + b^2} & \dfrac{-2ac}{a^2 + b^2} \\[2mm] \dfrac{-2ab}{a^2 + b^2} & \dfrac{(a^2 - b^2)}{a^2 + b^2} & \dfrac{-2bc}{a^2 + b^2} \\[2mm] 0 & 0 & 1 \end{bmatrix}, \quad \text{where } b \neq 0$$

A demonstration of this result involves a considerable amount of matrix algebra. To facilitate the demonstration, the necessary TI-92 matrix files are available in CD-ROM Activity cd4_5_1.

Example
4.5.1

Find the transformation matrix for a reflection in the line $[-1\ 3\ 3]$. Letting $a = -1, b = 3$, and $c = 3$, the transformation matrix simplifies to

$$R_1 = \begin{bmatrix} .8 & .6 & .6 \\ .6 & -.8 & -1.8 \\ 0 & 0 & 1 \end{bmatrix}$$

Determining the Matrix of a Linear Transformation Given an Object Set and an Image Set

A related task is that of determining the matrix of a linear transformation given an object set and its image under the transformation. That is, given a point set X and its image X', find the matrix A such that $AX = X'$. In addressing this task, the first issue to be considered is, "How many points are necessary?" Thinking analytically, this question may be rephrased as: Given the values of x, y, x', and y', are the values a, b, c, d, e, and f uniquely determined by the matrix equation

$$\begin{bmatrix} a & b & e \\ c & d & f \\ 0 & 0 & 1 \end{bmatrix} \begin{bmatrix} x \\ y \\ 1 \end{bmatrix} = \begin{bmatrix} x' \\ y' \\ 1 \end{bmatrix}$$

For instance, if we let $(x, y, 1) = (1, 1, 1)$ and $(x', y', 1) = (2, 1, 1)$, this equation yields the algebraic equations $a + b + e = 2$ and $c + d + f = 1$. Clearly, there are many choices of values for a, b, c, d, e, and f that satisfy these equations. One could imagine a translation T

$$\begin{bmatrix} 1 & 0 & 1 \\ 0 & 1 & 0 \\ 0 & 0 & 1 \end{bmatrix}$$

mapping $(1, 1, 1)$ onto $(2, 1, 1)$. Alternatively, the transformation given by the matrix

$$\begin{bmatrix} 1 & 1 & 1 \\ 0 & 1 & 0 \\ 0 & 0 & 1 \end{bmatrix}$$

will achieve the same result. So, one point and its image are insufficient to uniquely determine a linear transformation.

Might two points suffice? The matrix equation for two points and their images is

$$\begin{bmatrix} a & b & e \\ c & d & f \\ 0 & 0 & 1 \end{bmatrix} \begin{bmatrix} x_1 & x_2 \\ y_1 & y_2 \\ 1 & 1 \end{bmatrix} = \begin{bmatrix} x_1' & x_2' \\ y_1' & y_2' \\ 1 & 1 \end{bmatrix}$$

To explore this possibility, let the transformation map the point $(1, 1, 1)$ onto $(2, 1, 1)$ and the point $(0, 0, 1)$ onto $(1, 1, 1)$. These assumptions lead to observations that $a + b = 1$ and $c + d = 0$. Since there are an infinite number of solutions to these equations, using two points does not uniquely determine a linear equation.

What about using three points? Under this assumption, the matrix equation becomes

$$\begin{bmatrix} a & b & e \\ c & d & f \\ 0 & 0 & 1 \end{bmatrix} \begin{bmatrix} x_1 & x_2 & x_3 \\ y_1 & y_2 & y_3 \\ 1 & 1 & 1 \end{bmatrix} = \begin{bmatrix} x_1' & x_2' & x_3' \\ y_1' & y_2' & y_3' \\ 1 & 1 & 1 \end{bmatrix}$$

Matrix equations of this form, $AX = X'$, may be solved for A as long as $|X| \neq 0$. $|X| \neq 0$ whenever the three points are not collinear. When this requirement is satisfied, the solution may be written as $A = X'X^{-1}$.

Example 4.5.2   Find the matrix of the linear transformation T mapping the point set

$$X = \begin{bmatrix} 1 & 0 & -2 \\ 1 & 0 & 3 \\ 1 & 1 & 1 \end{bmatrix}$$

onto the point set

$$X' = \begin{bmatrix} -1 & 0 & -3 \\ 0 & 0 & 1 \\ 1 & 1 & 1 \end{bmatrix}$$

The first issue is to determine whether $|X| \neq 0$. A quick check will verify that the determinant is equal to -5. The next step is to determine the inverse of the matrix

$$X = \begin{bmatrix} 1 & 0 & -2 \\ 1 & 0 & 3 \\ 1 & 1 & 1 \end{bmatrix}$$

The inverse of matrix X is

$$X^{-1} = \begin{bmatrix} \dfrac{3}{5} & \dfrac{2}{5} & 0 \\ \dfrac{-2}{5} & \dfrac{-3}{5} & 1 \\ \dfrac{-1}{5} & \dfrac{1}{5} & 0 \end{bmatrix}$$

The unknown transformation is given by the expression $A = X'X^{-1}$, or

$$A = \begin{bmatrix} -1 & 0 & -3 \\ 0 & 0 & 1 \\ 1 & 1 & 1 \end{bmatrix} \begin{bmatrix} \dfrac{3}{5} & \dfrac{2}{5} & 0 \\ \dfrac{-2}{5} & \dfrac{-3}{5} & 1 \\ \dfrac{-1}{5} & \dfrac{1}{5} & 0 \end{bmatrix} = \begin{bmatrix} 0 & -1 & 0 \\ \dfrac{-1}{5} & \dfrac{1}{5} & 0 \\ 0 & 0 & 1 \end{bmatrix}$$

Invariant Lines and Eigenvectors

In sections 4.3 and 4.4, the equation $uA^{-1} = ku'$ was modified to $uA^{-1} = ku$ when searching for invariant lines under a linear transformation A. We now explore the relationship between the equation $uA^{-1} = ku'$ and another famous mathematical relationship, the eigen, or characteristic, equation of a matrix. The eigen equation is often written as $uT = \lambda u$, where T is a matrix, $\lambda \in R$, $u \in V$, and V is a vector space.

For a given matrix T, pairs of λ and u satisfying this equation are called eigenvalues and eigenvectors, respectively. Interpreting this relationship geometrically, an eigenvector u is rescaled by a factor λ, undergoing only a change in length and/or a reversal of direction. For this to happen, both u and its image must lie in the same line, l. As a result, the line l must be invariant under T. This suggests that standard procedures for determining eigenvalues and eigenvectors may be applied to the task of determining invariant lines under a given linear transformation. Comparing the line equation $uA^{-1} = ku$ and the eigen equation $uT = \lambda u$, it is apparent that the values obtained for k in the examples in sections 4.3–4.4 were, in fact, the eigenvalues associated with the transformation matrix A^{-1}.

The eigen equation $uT = \lambda u$ may be rewritten as $uT - \lambda u = 0$. Factoring out the vector u and multiplying λ by the identify matrix yields $u(T - \lambda I) = 0$. In matrix form, this is

$$\begin{bmatrix} u_1 & u_2 & u_3 \end{bmatrix} \begin{bmatrix} t_{11} - \lambda & t_{12} & t_{13} \\ t_{21} & t_{22} - \lambda & t_{23} \\ 0 & 0 & 1 - \lambda \end{bmatrix} = 0$$

It can be shown that this equation has nontrivial solutions when

$$\begin{vmatrix} t_{11} - \lambda & t_{12} & t_{13} \\ t_{21} & t_{22} - \lambda & t_{23} \\ 0 & 0 & 1 - \lambda \end{vmatrix} = 0$$

This determinant may be simplified and solved for λ. Once the eigenvalues have been determined, the eigenvectors are obtained by substituting these values into the eigen equation. These eigenvectors may be used to identify invariant lines. The advantage of making this association goes beyond mathematical appreciation, however. Many mathematical technologies such as the TI-92 calculator, Maple, and Matlab find eigenvalues and eigenvectors automatically.

Example 4.5.3   Find the eigenvalues and eigenvectors associated with the linear transformation

$$\begin{bmatrix} 1 & 0 & 0 \\ 0 & -1 & 0 \\ 0 & 0 & 1 \end{bmatrix}$$

The eigenvalues associated with this matrix are specified by the equation

$$\begin{vmatrix} 1 - \lambda & 0 & 0 \\ 0 & -1 - \lambda & 0 \\ 0 & 0 & 1 - \lambda \end{vmatrix} = 0$$

should be:
$(1+\lambda)(1-\lambda)^2 \leftarrow$ Expanding and simplifying terms yields the algebraic equation $(1 + \lambda)(1 - \lambda)2 = 0$. Solving this equation yields the values $\lambda = \pm 1$. Each of these values is substituted back into the eigen equation

$$[u_1 \quad u_2 \quad u_3] \begin{bmatrix} 1 & 0 & 0 \\ 0 & -1 & 0 \\ 0 & 0 & 1 \end{bmatrix} = \lambda[u_1 \quad u_2 \quad u_3]$$

If $\lambda = 1$, this equation yields $u_1 = u_1$, $-u_2 = u_2$, and $u_3 = u_3$. The interpretation of these results is that $u_2 = 0$, leaving u_1 and u_3 unconstrained. The eigenvector associated with this eigenvalue is of the form $[u_1 \quad 0 \quad u_3]$, or $x = u_3/u_1$. All such lines are perpendicular to the x-axis. If $\lambda = -1$, this equation yields $u_1 = -u_1$, $-u_2 = -u_2$, and $u_3 = -u_3$. The interpretation of these results is that $u_1 = u_3 = 0$, leaving u_2 unconstrained. The eigenvector is associated with all lines of the form $[0 \quad u_2 \quad 0]$, or $y = 0$. This is the x-axis. It is left as an exercise to show that the only possible eigenvalues for an isometry are 1 and -1.

Summary

Matrices for rotations about turn points other than the origin and reflections in lines other than the x-axis are easily determined by composing simpler matrices. The general strategy is to move the turn center at the origin or the reflecting line to the *x*-axis, perform the intended transformation, then move the turn center or reflecting line back to its original position. Composing these separate matrices by multiplying them from left to right (i.e., $T^{-1}RTX = X'$ for a rotation or $T^{-1}R^{-1}FRTX = X'$ for a reflection) produces the desired result. Finally, a second method of identifying invariant lines was introduced by associating them with the eigenvectors of the transformation matrix.

Rotation of a Point about a Turn Center Other Than the Origin

Q: What is the image of the point $(x, y, 1)$ under a rotation of $\theta°$ about the point $(e, f, 1)$?

A: A point $(x', y', 1)$

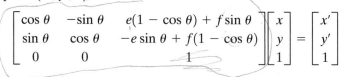

$$\begin{bmatrix} \cos\theta & -\sin\theta & e(1 - \cos\theta) + f\sin\theta \\ \sin\theta & \cos\theta & -e\sin\theta + f(1 - \cos\theta) \\ 0 & 0 & 1 \end{bmatrix} \begin{bmatrix} x \\ y \\ 1 \end{bmatrix} = \begin{bmatrix} x' \\ y' \\ 1 \end{bmatrix}$$

Reflection of a Point in a Line Other Than the *X*-Axis

Q: What is the image of the point $(x, y, 1)$ under a reflection in the line $\begin{bmatrix} a & b & c \end{bmatrix}$?

A: A point $(x', y', 1)$

$$\begin{bmatrix} \dfrac{-(a^2 - b^2)}{a^2 + b^2} & \dfrac{-2ab}{a^2 + b^2} & \dfrac{-2ac}{a^2 + b^2} \\ \dfrac{-2ab}{a^2 + b^2} & \dfrac{(a^2 - b^2)}{a^2 + b^2} & \dfrac{-2bc}{a^2 + b^2} \\ 0 & 0 & 1 \end{bmatrix} \begin{bmatrix} x \\ y \\ 1 \end{bmatrix} = \begin{bmatrix} x' \\ y' \\ 1 \end{bmatrix}, \quad \text{where } b \neq 0$$

Determine the Matrix of a Linear Transformation

Q: What is the matrix of a linear transformation given an object set X and its image X' under the given transformation.

A: The matrix A

$$A = \begin{bmatrix} a & b & e \\ c & d & f \\ 0 & 0 & 1 \end{bmatrix} = \begin{bmatrix} x'_1 & x'_2 & x'_3 \\ y'_1 & y'_2 & y'_3 \\ 1 & 1 & 1 \end{bmatrix} \begin{bmatrix} x_1 & x_2 & x_3 \\ y_1 & y_2 & y_3 \\ 1 & 1 & 1 \end{bmatrix}^{-1}$$

Exercises

1. Using geometric thinking, sketch the composition of two translations.
2. Using geometric thinking, sketch the composition of a reflection and a translation.
3. Using geometric thinking, sketch the composition of two reflections.
4. Write transformation matrices for 90 degree rotations about each of the following turn centers: $(1, 1, 1)$, $(-1, 1, 1)$, $(-1, -1, 1)$, and $(1, -1, 1)$. Discuss their similarities and differences.

5. Write transformation matrices for reflections in each of the following lines: $y = x$, $y = x + 1$, $y = x + 2$, $y = x - 1$, $y = x - 2$. Discuss their similarities and differences.
6. Find the composition $C = TR$ and determine the image of the point $X(1, 5, 1)$ under C where

$$T = \begin{bmatrix} 1 & 0 & -2 \\ 0 & 1 & 1 \\ 0 & 0 & 1 \end{bmatrix} \quad \text{and} \quad R = \begin{bmatrix} 0 & -1 & 0 \\ 1 & 0 & 0 \\ 0 & 0 & 1 \end{bmatrix}$$

7. Find the composition $C = RT$ and determine the image of the point $X(1, 5, 1)$ under C using the same matrices for T and R as the previous exercise. Compare the results obtained in exercises 6 and 7.
8. Using a composition of matrices, find the image of $X(1, 5, 1)$ in a rotation of an angle θ about the point $C(-1, -1, 1)$.
9. Using a composition of matrices, find the image of $X(1, 5, 1)$ in a reflection in the line $y = -5x + 4$.
10. Under an unknown linear transformation T, the point set

$$\begin{bmatrix} 0 & -1 & -3 \\ 0 & -1 & 0 \\ 1 & 1 & 1 \end{bmatrix}$$

is mapped onto the point set

$$\begin{bmatrix} \sqrt{3} & \dfrac{3\sqrt{3} + 1}{2} & \sqrt{3} + \dfrac{3}{2} \\ 1 & \dfrac{\sqrt{3} + 1}{2} & \dfrac{3\sqrt{3} + 2}{2} \\ 1 & 1 & 1 \end{bmatrix}$$

a) Find a transformation matrix for T and identify all points and lines that are invariant under the transformation. Give a geometric description of the transformation.
b) Find a composition of transformations equivalent to the transformation T. Write the composition in matrix form.
11. Given a line l and a point P on l, find a transformation matrix for determining a line m perpendicular to l at P. Give an example.
12. Given two intersecting lines l and m, find a transformation matrix for repositioning their point of intersection at the origin with line l coincident with the x-axis.
13. Given a translation T and reflection R, where

$$T = \begin{bmatrix} 1 & 0 & 0 \\ 0 & 1 & -1 \\ 0 & 0 & 1 \end{bmatrix} \quad \text{and} \quad R = \begin{bmatrix} 1 & 0 & 0 \\ 0 & -1 & 0 \\ 0 & 0 & 1 \end{bmatrix}$$

a) Find any invariant lines associated with
   i) T
   ii) R
   iii) TR
   iv) RT
b) Find any eigenvectors associated with
   i) T
   ii) R
   iii) TR
   iv) RT
c) Compare the answers obtained in parts a) and b). Are they the same or different? Why?

14. Use both geometric thinking and matrix methods to explain the existence and nature of compositions of linear transformations. How does each approach complement the other?

15. What is the relationship between the eigenvalues and eigenvectors of individual isometries and their composition?

16. Prove Theorem 4.5.1.

17. Prove Theorem 4.5.2.

18. Prove that the only possible eigenvalues for an isometry are 1 and -1.

4.6 Other Linear Transformations

In this section, you will . . .

Use both geometric and matrix methods to:
- **Represent dilations, strains, shears, and affinities;**
- **Identify invariant points and lines associated with these transformations;**
- **Determine whether distance and area are preserved under these transformations.**

All of the linear transformations considered so far have been isometries. As such, they preserve both the size and shape of objects. Other sorts of linear transformations are possible. Each of the transformations discussed in this section is not an isometry.

Representing Dilations

Definition 4.6.1

A *dilation D* with nonzero ratio r is a one-to-one linear transformation of V onto itself such that for each pair of points P and Q separated by a directed distance d, their images under the dilation $D(P)$ and $D(Q)$ are separated by the directed distance $r*d$.

A dilation D has one of the following matrix representations:

$$\begin{bmatrix} a & b & e \\ -b & a & f \\ 0 & 0 & 1 \end{bmatrix} \quad \text{or} \quad \begin{bmatrix} a & b & e \\ b & -a & f \\ 0 & 0 & 1 \end{bmatrix}, \text{where } a^2 + b^2 = r^2.$$

Dilations with matrices of the first sort are called *direct*. Dilations with matrices of the second sort are called *indirect*. Dilation D is a special case of the direct dilations, having center $(0, 0, 1)$ and ratio r.

$$D = \begin{bmatrix} r & 0 & 0 \\ 0 & r & 0 \\ 0 & 0 & 1 \end{bmatrix}$$

Geometrically, dilations are like photographic enlargements or reductions (See Figure 4.6.1). The object set, $\triangle ABC$, and its image set, $\triangle A'B'C'$, are similar, having equal angles and corresponding sides in proportion. For this reason, dilations are also known as similarity transformations.

Theorem 4.6.1

The transformation matrix D for a direct dilation with center $(h, k, 1)$ and ratio r is

$$D = \begin{bmatrix} r & 0 & h(1 - r) \\ 0 & r & k(1 - r) \\ 0 & 0 & 1 \end{bmatrix}$$

The proof of this theorem is left as an exercise.

Example 4.6.1

Write the transformation matrix for a dilation with center $(2, -3, 1)$ and ratio 2 and find the image of the origin under the dilation.

The matrix equation and solution required is

$$\begin{bmatrix} 2 & 0 & -2 \\ 0 & 2 & 3 \\ 0 & 0 & 1 \end{bmatrix} \begin{bmatrix} 0 \\ 0 \\ 1 \end{bmatrix} = \begin{bmatrix} -2 \\ 3 \\ 1 \end{bmatrix}$$

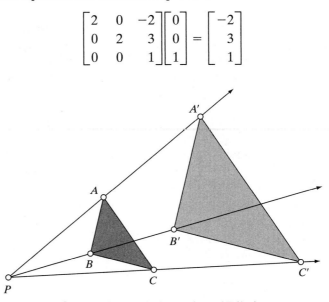

Figure 4.6.1 Transformation of Dilation

Add a third triangle to Figure 4.6.1 representing a second dilation of $\triangle ABC$ having the same center but a different ratio.

Invariance Under Dilation

The existence of invariant points and lines under dilations may be explored using both geometric and analytic approaches. Thinking geometrically, unless $r = 1$, the only invariant point associated with a dilation D is the center of the dilation, $P(h, k, 1)$. All other points move away from or toward P, depending on whether r is greater than or less than 1, along rays originating at P. This suggests that all lines containing P are invariant under the dilation. So, a geometric analysis suggests that there should be a single invariant point associated with each dilation, the center, and infinitely many invariant lines, each passing through the center of the dilation.

These observations may be verified using an analytic approach focused on the transformation matrix, D. Invariant points may be identified using the matrix equation

$$\begin{bmatrix} r & 0 & h(1-r) \\ 0 & r & k(1-r) \\ 0 & 0 & 1 \end{bmatrix} \begin{bmatrix} x \\ y \\ 1 \end{bmatrix} = \begin{bmatrix} x \\ y \\ 1 \end{bmatrix}$$

This approach leads to the algebraic equations $rx + h(1 - r) = x$ and $ry + (1 - r) = y$. Solving these equations for x and y yields $x = h$ and $y = k$, the center of the dilation. Before considering the existence of invariant lines under dilation, the following theorem is necessary.

**Theorem
4.6.2**

The inverse matrix for a dilation D with center $(h, k, 1)$ and ratio r is a dilation D^{-1} with center $(h, k, 1)$ and ratio $1/r$. The matrix representation for D^{-1} is

$$D^{-1} = \begin{bmatrix} \dfrac{1}{r} & 0 & h\left(1 - \dfrac{1}{r}\right) \\ 0 & \dfrac{1}{r} & k\left(1 - \dfrac{1}{r}\right) \\ 0 & 0 & 1 \end{bmatrix}$$

The proof of this theorem is left as an exercise.

**Example
4.6.2**

Write the inverse of the matrix used in Example 4.6.1, then show that the product of the two matrices is the identity matrix.

$$\begin{bmatrix} \dfrac{1}{2} & 0 & 1 \\ 0 & \dfrac{1}{2} & \dfrac{-3}{2} \\ 0 & 0 & 1 \end{bmatrix} \begin{bmatrix} 2 & 0 & -2 \\ 0 & 2 & 3 \\ 0 & 0 & 1 \end{bmatrix} = \begin{bmatrix} 1 & 0 & 0 \\ 0 & 1 & 0 \\ 0 & 0 & 1 \end{bmatrix}$$

Returning to the question of invariant lines, a geometric analysis suggests that all lines through the center of the dilation are invariant. This may now be confirmed using an analytic approach. First, it is clear that every line through the center of the dilation contains the point $P(h, k, 1)$. All lines incident with the point $P(h, k, 1)$ have equations of the form

$$\begin{bmatrix} u_1 & u_2 & u_3 \end{bmatrix} \begin{bmatrix} h \\ k \\ 1 \end{bmatrix} = 0$$

Since the parameters u_1 and u_2 are associated with the slopes of these lines, and since lines of all possible slopes pass through P, these parameters must be unrestricted. On the other hand, u_3 must be dependent on u_1 and u_2 to satisfy the incidence equation above. Specifically, $u_3 = -hu_1 - ku_2$ and all invariant lines under the dilation have form $\begin{bmatrix} u_1 & u_2 & (-hu_1 - ku_2) \end{bmatrix}$. This fact may be verified using the line equation $uD^{-1} = ku$, where k is a constant not necessarily the same as the y-coefficient of the center of the dilation. Simplification leads to

$$\begin{bmatrix} u_1 & u_2 & (-hu_1 - ku_2) \end{bmatrix} \begin{bmatrix} \dfrac{1}{r} & 0 & h\left(1 - \dfrac{1}{r}\right) \\ 0 & \dfrac{1}{r} & k\left(1 - \dfrac{1}{r}\right) \\ 0 & 0 & 1 \end{bmatrix} = \left(\dfrac{1}{r}\right) \begin{bmatrix} u_1 & u_2 & (-hu_1 - ku_2) \end{bmatrix}$$

So, the analytic approach confirms the insights of the geometric approach. It is interesting to note that the constant k associated with the line equation turns out to be the same as the dilation ratio, $1/r$.

Representing Strains

A strain S with ratio r and axis $[0 \quad 1 \quad 0]$, also known as the x-axis, has matrix representation

$$S = \begin{bmatrix} 1 & 0 & 0 \\ 0 & r & 0 \\ 0 & 0 & 1 \end{bmatrix}$$

Conceptually, strains stretch or compress objects (See Figure 4.6.2). Under strains, distance is rescaled in one dimension (perpendicular to the strain axis) while distance in the other dimension (parallel to the strain axis) remains unchanged.

Theorem 4.6.3

The inverse matrix for a strain S with axis $[0 \quad 1 \quad 0]$ and ratio r is

$$S^{-1} = \begin{bmatrix} 1 & 0 & 0 \\ 0 & \dfrac{1}{r} & 0 \\ 0 & 0 & 1 \end{bmatrix}$$

The proof of this theorem is left as an exercise.

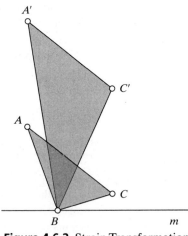

Figure 4.6.2 Strain Transformation

Add a third triangle to Figure 4.6.2 representing a second strain of $\triangle ABC$ having the same axis but a different ratio.

Transformation matrices for strains with axes other than $[0 \quad 1 \quad 0]$ may be developed using a composition of matrices as described in Section 4.5. The following example illustrates a composition of matrices for a strain in the line $y = x$, or $[1 \quad -1 \quad 0]$.

Example 4.6.3

Find a matrix equation for determining the image of $X(x, y, 1)$ in the strain axis $[1 \quad -1 \quad 0]$.

While there are several approaches to this task, for this example the following strategy is employed:

- Reposition the strain axis by rotating through an angle of $-45°$ about the origin;
- Perform the strain in the line $\begin{bmatrix} 0 & 1 & 0 \end{bmatrix}$;
- Return the strain axis to its original position by rotating through an angle of $45°$ about the origin.

Implementing this strategy using the respective matrices yields the following matrix equation:

$$\begin{bmatrix} \cos 45 & -\sin 45 & 0 \\ \sin 45 & \cos 45 & 0 \\ 0 & 0 & 1 \end{bmatrix} \begin{bmatrix} 1 & 0 & 0 \\ 0 & r & 0 \\ 0 & 0 & 1 \end{bmatrix} \begin{bmatrix} \cos(-45) & -\sin(-45) & 0 \\ \sin(-45) & \cos(-45) & 0 \\ 0 & 0 & 1 \end{bmatrix} \begin{bmatrix} x \\ y \\ 1 \end{bmatrix} = \begin{bmatrix} x' \\ y' \\ 1 \end{bmatrix}, \quad \text{or}$$

$$\begin{bmatrix} \dfrac{1+r}{2} & \dfrac{1-r}{2} & 0 \\ \dfrac{1-r}{2} & \dfrac{1+r}{2} & 0 \\ 0 & 0 & 1 \end{bmatrix} \begin{bmatrix} x \\ y \\ 1 \end{bmatrix} = \begin{bmatrix} x' \\ y' \\ 1 \end{bmatrix}$$

Invariance Under Strain

The existence of invariant points and lines may be approached both geometrically and analytically. Thinking geometrically, the strain will have no effect on points lying on the strain axis, so all points on the strain axis are invariant and the line containing the strain axis is point-wise invariant. Under the strain, all other points are moved along lines perpendicular to the strain axis. So lines perpendicular to the strain axis are also invariant. As expected, an analytic approach yields the same results.

Example 4.6.4 Using an analytic approach find all invariant points under the strain in Example 4.6.2.

Expanding the matrix equation

$$\begin{bmatrix} \dfrac{1+r}{2} & \dfrac{1-r}{2} & 0 \\ \dfrac{1-r}{2} & \dfrac{1+r}{2} & 0 \\ 0 & 0 & 1 \end{bmatrix} \begin{bmatrix} x \\ y \\ 1 \end{bmatrix} = \begin{bmatrix} x' \\ y' \\ 1 \end{bmatrix}$$

yields two algebraic equations, $[(1 + r)/2]x + [(1 - r)/2]y = x$ and $[(1 - r)/2]x + [(1 + r)/2]y = y$. When simplified, both of these equations reduce to $y = x$, or $\begin{bmatrix} 1 & -1 & 0 \end{bmatrix}$ in matrix form. So, all points on the strain axis are invariant. This makes the strain axis point-wise invariant. An analytic approach to the identification of invariant lines under a strain is left to the exercises.

Representing Shears

A shear S with axis $\begin{bmatrix} 0 & 1 & 0 \end{bmatrix}$ and ratio r has matrix representation

$$S = \begin{bmatrix} 1 & r & 0 \\ 0 & 1 & 0 \\ 0 & 0 & 1 \end{bmatrix}$$

Conceptually, a shear holds an axis point-wise invariant and slides all other points parallel to the axis (See Figure 4.6.3). This motion may be demonstrated in three dimensions using a deck of playing cards. A deck set squarely on the table may be viewed as a right parallelopiped. If one edge of the deck is gradually shifted from a vertical to an angled relationship with the table, the deck assumes the shape of an oblique parallelopiped. This motion, from a right to an oblique parallelopiped, is a shear in three dimensions.

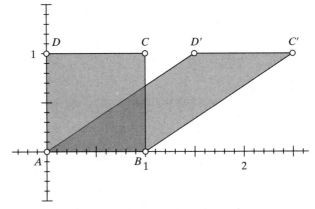

Figure 4.6.3 Shear Transformation

Theorem 4.6.4 The inverse matrix for a shear S with axis $\begin{bmatrix} 0 & 1 & 0 \end{bmatrix}$ and ratio r is

$$S^{-1} = \begin{bmatrix} 1 & -r & 0 \\ 0 & 1 & 0 \\ 0 & 0 & 1 \end{bmatrix}$$

The proof of this theorem is left as an exercise.

Transformation matrices for shears with axes other than $\begin{bmatrix} 0 & 1 & 0 \end{bmatrix}$ may be developed using a composition of matrices as described in Section 4.5. The following example illustrates a composition of matrices for a shear in the line $y = x$, or $\begin{bmatrix} 1 & -1 & 0 \end{bmatrix}$.

Example 4.6.5 Find a matrix equation for finding the image of $X(x, y, 1)$ in the shear axis $\begin{bmatrix} 1 & -1 & 0 \end{bmatrix}$.

Following the same strategy used in Example 4.6.3,

- Reposition the shear axis by rotating through an angle of $-45°$ about the origin;

- Perform the shear in the line $[0 \quad 1 \quad 0]$;
- Return the shear axis to its original position by rotating through an angle of 45° about the origin.

Implementing this strategy using the respective matrices yields the following matrix equation:

$$\begin{bmatrix} \cos 45 & -\sin 45 & 0 \\ \sin 45 & \cos 45 & 0 \\ 0 & 0 & 1 \end{bmatrix} \begin{bmatrix} 1 & -r & 0 \\ 0 & 1 & 0 \\ 0 & 0 & 1 \end{bmatrix} \begin{bmatrix} \cos(-45) & -\sin(-45) & 0 \\ \sin(-45) & \cos(-45) & 0 \\ 0 & 0 & 1 \end{bmatrix} \begin{bmatrix} x \\ y \\ 1 \end{bmatrix} = \begin{bmatrix} x' \\ y' \\ 1 \end{bmatrix}, \quad \text{or}$$

$$\begin{bmatrix} \dfrac{2+r}{2} & \dfrac{-r}{2} & 0 \\ \dfrac{r}{2} & \dfrac{2-r}{2} & 0 \\ 0 & 0 & 1 \end{bmatrix} \begin{bmatrix} x \\ y \\ 1 \end{bmatrix} = \begin{bmatrix} x' \\ y' \\ 1 \end{bmatrix}$$

Using geometric and analytic approaches similar to those taken when investigating invariance under dilations and strains, it is clear that the only invariant points under a shear are points on the shear axis and that all lines parallel to the shear axis are invariant.

Representing Affinities

Affine transformations, or affinities, are general invertible transformations (see Figure 4.6.4).

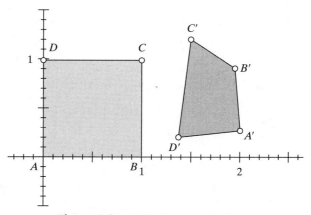

Figure 4.6.4 Affine Transformation

The general matrix for affine transformations is

$$T = \begin{bmatrix} a & b & e \\ c & d & f \\ 0 & 0 & 1 \end{bmatrix}$$

General Features of Affinities

Isometries, dilations, strains, and shears are affine transformations having special characteristics. The following characteristics are common to all affine transformations and therefore to each of the linear transformations discussed in this chapter.

Theorem 4.6.5

Given any affine transformation T, if m and n are parallel lines, their images under T are also parallel.

Assume that $u[u_1 \quad u_2 \quad u_3]$ and $v[v_1 \quad v_2 \quad v_3]$ are parallel. Using the line equation $uT^{-1} = ku'$ with

$$T = \begin{bmatrix} a & b & e \\ c & d & f \\ 0 & 0 & 1 \end{bmatrix} \quad \text{and} \quad T^{-1} = \begin{bmatrix} g & h & m \\ i & j & l \\ 0 & 0 & 1 \end{bmatrix}$$

we may write the matrix equations

$$[u_1 \quad u_2 \quad u_3] \begin{bmatrix} g & h & m \\ i & j & l \\ 0 & 0 & 1 \end{bmatrix} = k[u_1' \quad u_2' \quad u_3'] \quad \text{and}$$

$$[v_1 \quad v_2 \quad v_3] \begin{bmatrix} g & h & m \\ i & j & l \\ 0 & 0 & 1 \end{bmatrix} = k[v_1' \quad v_2' \quad v_3']$$

Expanding these matrix equations yields

$$[(gu_1 + iu_2)(hu_1 + ju_2)(mu_1 + lu_2 + u_3)] = [ku_1' \quad ku_2' \quad ku_3'] \quad \text{and}$$

$$[(gv_1 + iv_2)(hv_1 + jv_2)(mv_1 + lv_2 + v_3)] = [kv_1' \quad kv_2' \quad kv_3']$$

Because u and v are parallel, there is a nonzero real number t such that $v_1 = tu_1$ and $v_2 = tu_2$. Then

$$[(gv_1 + iv_2)(hv_1 + jv_2)(mv_1 + lv_2 + v_3)] = [kv_1' \quad kv_2' \quad kv_3']$$

may be written as

$$[(gtu_1 + itu_2)(htu_1 + jtu_2)(mtu_1 + ltu_2 + v_3)] = [tku_1' \quad tku_2' \quad kv_3'].$$

As a result, the images v' and u' are parallel.

Theorem 4.6.6

Given any affine transformation T, if $A, B,$ and C are collinear points with B between A and C, and if $d(A,B)/d(A,C) = r$, then $d(T(A),T(B))/d(T(A),T(C)) = r$, where r is the segment division ratio.

CD 4.6.1

The proof this theorem involves a great deal of symbol manipulation. Students wishing to see a demonstration are invited to transfer the necessary TI-92 files from cd4_6_1 and work through the steps found in that laboratory activity.

**Example
4.6.6**

Show that the segment ratios for $X(0, 0, 1)$, $Y(1, 0, 1)$, and $Z(3, 0, 1)$ and their images under the affine transformation T are the same, where

$$T = \begin{bmatrix} 1 & 1 & 0 \\ 0 & 2 & 1 \\ 0 & 0 & 1 \end{bmatrix}$$

The segment ratio for points X, Y, and Z is $1/3$. The images of these points under T are given by

$$\begin{bmatrix} 1 & 1 & 0 \\ 0 & 2 & 1 \\ 0 & 0 & 1 \end{bmatrix} \begin{bmatrix} 0 & 1 & 3 \\ 0 & 0 & 0 \\ 1 & 1 & 1 \end{bmatrix} = \begin{bmatrix} 0 & 1 & 3 \\ 1 & 1 & 1 \\ 1 & 1 & 1 \end{bmatrix}$$

Using the distance formula, $d(T(X),T(Y)) = \sqrt{(0-1)^2 + (1-1)^2} = 1$ and $d(T(X),T(Z)) = \sqrt{(0-3)^2 + (1-1)^2} = 3$. The segment ratio for these points is also $1/3$. So, the segment ratio is preserved under the transformation.

**Theorem
4.6.7**

Affine transformations preserve betweeness of points.

Given collinear points A, B, and C, with B between A and C, $d(A, B) + d(B, C) = d(A, C)$ by the definition of betweeness. The segment ratio associated with these points is $k = d(A, B)/d(A, C) < 1$. Under an affine transformation T, $T(A)$, $T(B)$, and $T(C)$ are also collinear. T also preserves segment ratio, so $k = d(T(A), T(B))/d(T(A), T(C)) < 1$. Then $T(B)$ is between $T(A)$ and $T(C)$.

**Theorem
4.6.8**

Given affine transformation T, if m is the midpoint of segment AB, then $T(m)$ is the midpoint of segment $T(A)T(B)$.

This theorem is a special case of Theorem 4.6.6 in which the segment ratio is $1/2$.

Summary

In addition to the isometries, dilations, strains, and shears are commonly used linear transformations. Each of these transformations has one or more invariant points and an infinite number of invariant lines. Affinities are the most general of linear transformations, preserving collinearity, concurrence, betweeness, and the segment ratio. Every affinity is equivalent to a composition of a shear, a strain, and a direct similarity.

Dilation Transformation
Q: What is the image of the point $(x, y, 1)$ under a dilation about the point $(h, k, 1)$ with ratio r?

A: A point $(x', y', 1)$

$$\begin{bmatrix} r & 0 & h(1-r) \\ 0 & r & k(1-r) \\ 0 & 0 & 1 \end{bmatrix} \begin{bmatrix} x \\ y \\ 1 \end{bmatrix} = \begin{bmatrix} x' \\ y' \\ 1 \end{bmatrix}$$

Strain Transformation
Q: What is the image of the point $(x, y, 1)$ under a strain with ratio r in the axis $\begin{bmatrix} 0 & 1 & 0 \end{bmatrix}$?
A: A point $(x', y', 1)$

$$\begin{bmatrix} 1 & 0 & 0 \\ 0 & r & 0 \\ 0 & 0 & 1 \end{bmatrix} \begin{bmatrix} x \\ y \\ 1 \end{bmatrix} = \begin{bmatrix} x' \\ y' \\ 1 \end{bmatrix}$$

Shear Transformation
Q: What is the image of the point $(x, y, 1)$ under a shear with ratio r in the axis $\begin{bmatrix} 0 & 1 & 0 \end{bmatrix}$?
A: A point $(x', y', 1)$

$$\begin{bmatrix} 1 & r & 0 \\ 0 & 1 & 0 \\ 0 & 0 & 1 \end{bmatrix} \begin{bmatrix} x \\ y \\ 1 \end{bmatrix} = \begin{bmatrix} x' \\ y' \\ 1 \end{bmatrix}$$

Affine Transformation
Q: What is the image of the point $(x, y, 1)$ under an affinity?
A: A point $(x', y', 1)$

$$\begin{bmatrix} a & b & e \\ c & d & f \\ 0 & 0 & 1 \end{bmatrix} \begin{bmatrix} x \\ y \\ 1 \end{bmatrix} = \begin{bmatrix} x' \\ y' \\ 1 \end{bmatrix}$$

Exercises

1. Sketch a dilation with center $(1, 1, 1)$ and ratio $r = 1.5$.
2. Sketch a dilation with center $(0, 0, 1)$ and $r = .5$.
3. Sketch a shear with axis $\begin{bmatrix} 1 & 0 & 0 \end{bmatrix}$ and ratio $r = 2$.
4. Sketch a strain with axis $\begin{bmatrix} 1 & 1 & 0 \end{bmatrix}$ and ratio $r = 2$.
5. A triangle has vertices $(0, 0, 1)$, $(2, 3, 1)$, and $(5, -1, 1)$. Determine the images of these points and plot each image triangle under each of the following transformations.
   a) A dilation with center $(1, 2, 1)$ and $r = 2$
   b) A strain with axis $\begin{bmatrix} 0 & 1 & -2 \end{bmatrix}$ and $r = 2$
   c) A shear with axis $\begin{bmatrix} 1 & 0 & 1 \end{bmatrix}$ and $r = 2$
6. Identify each of the following transformations and any invariant points and lines.

a) $\begin{bmatrix} 3 & 0 & -4 \\ 0 & 3 & -8 \\ 0 & 0 & 1 \end{bmatrix} \begin{bmatrix} x \\ y \\ 1 \end{bmatrix} = \begin{bmatrix} x' \\ y' \\ 1 \end{bmatrix}$ b) $\begin{bmatrix} 1 & 0 & 0 \\ 0 & 4 & -3 \\ 0 & 0 & 1 \end{bmatrix} \begin{bmatrix} x \\ y \\ 1 \end{bmatrix} = \begin{bmatrix} x' \\ y' \\ 1 \end{bmatrix}$

7. Find a composition of matrices involving a direct similarity, strain, and a shear equivalent to the affine transformation *T*.

$$T = \begin{bmatrix} 1 & 2 & 3 \\ 3 & 2 & 1 \\ 0 & 0 & 1 \end{bmatrix}$$

8. Affinities are the most general of the linear transformations. Both geometric and analytic approaches are useful in analyzing their invariant points and lines and determining their general features. Describe your preferred strategies for approaching this task.
9. Prove Theorem 4.6.1.
10. Prove Theorem 4.6.2.
11. Prove Theorem 4.6.3.
12. Prove Theorem 4.6.4.
13. Prove Theorem 4.6.6.

References and Suggested Reading

Cederberg, J.N. 1989. *A Course in modern geometries.* New York: Springer-Verlag.

Eddins, S.K., E.O. Maxwell, and F. Stanislaus. 1994. *Opportunities to become familiar with translations, rotations, and reflections.* Mathematics Teacher, 87, 177–181, 187–189.

Eddins, S.K., E.O. Maxwell , and F. Stanislaus. 1994. *Coordinate approaches to transformations utilizing matrices.* Mathematics Teacher, 87, 258–261, 268–270.

Edwards, L.D. 1991. "Children's Learning in a Computer Microworld for Transformation Geometry." *Journal for Research in Mathematics Education,* 22 (2, 122–137).

Jaime, A. and A. Gutiérrez. 1995. *Connecting Research to Teaching. Isometries as a link for different branches of mathematics or for mathematics and other sciences.* Mathematics Teacher, 88, 591–597.

Kazmierczak, M. *Matrix Calculator Applet.* Available on-line http://www.mkaz. com/math/matrix.html

Ludwig, H. *Geometry Bibliography: Tessellations.* Available on-line http://forum. swarthmore.edu/mathed/mtbib/tessellations.html

Ludwig, H. *Geometry Bibliography: Transformational Geometry.* Available on-line http://forum.swarthmore.edu/mathed/mtbib/transformations.html

National Council of Teachers of Mathematics 2000. *Principles and Standards for School Mathematics.* Available on-line http://standards-e.nctm.org/protoFINAL/ cover.html

O'Connor, J.J. and E.F. Robertson. 1999. *The MacTutor History of Mathematics archive.* School of Mathematics and Statistics, University of St. Andrews, Scotland. Available on-line:
Ames. Available on-line http://www-history.mcs.st-and.ac.uk/history/ Mathematicians/Ames.html

Cayley. Available on-line http://www-history.mcs.st-andrews.ac.uk/history/ Mathematicians/Cayley.html

Descartes. Available on-line http://www-history.mcs.st-andrews.ac.uk/history/ Mathematicians/Descartes.html

Fermat. Available on-line http://www-history.mcs.st-andrews.ac.uk/history/ Mathematicians/Fermat.html

Klein. Available on-line http://www-history.mcs.st-andrews.ac.uk/history/ Mathematicians/Klein.html

Leibnitz. Available on-line http://www-history.mcs.st-andrews.ac.uk/history/ Mathematicians/Leibniz.html

Newton. Available on-line http://www-history.mcs.st-andrews.ac.uk/history/ Mathematicians/Newton.html

Steve Okolica, S. and G. Macrina. 1992. *Moving transformation geometry ahead of deductive geometry.* Mathematics Teacher, 85, 716–719.

ScienceU. *Geometry Technologies, Inc.* Available on-line:

Isometries: *Reflection.* Available on-line http://www.ScienceU.com/library/ articles/ isometries/reflection.html

Isometries: *Rotation.* Available on-line http://www.ScienceU.com/library/ articles/ isometries/rotation.html

Isometries: *Translation.* Available on-line http://www.ScienceU.com/ library/articles/isometries/translation.html

Thomas, D. 1990. "Give meaning to matrices with MATLAB." *Journal of computers in mathematics and science teaching,* 9(3), 73–85.

Thomas, D. 1990. "Turtle tessellations." *Journal of computers in mathematics and science teaching,* 9(4), 25–34.

Chapter 5

Fractal Geometry

The next time you are at the grocery store, purchase a head of cauliflower. When you get home, cut the head in half and examine its internal structure (see Figure 5.1.1). Notice that the overall geometry of the cauliflower head is repeated on smaller and smaller scales as the cauliflower branches into smaller units. Each of the smaller units is similar in structure to the larger units. Because portions of the whole are similar in form to the whole cauliflower, the cauliflower is said to be "self-similar." Can you think of other vegetables that are self-similar? What about other objects in nature?

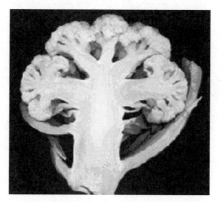

Figure 5.1.1 Cauliflower Head

Ignoring differences introduced by shading, Figure 5.1.2 is a purely geometric example of self-similarity: The smaller parts are similar to the whole.

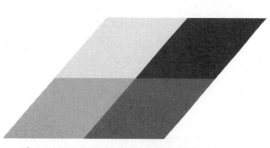

Figure 5.1.2 Self-similar Parallelograms

201

Definition
5.1.1

Any object composed of smaller models of itself is said to be *self-similar*.

Self-similar objects such as those seen in Figure 5.1.2 have been used for thousands of years in decorative applications of mathematics. In a clear break with traditional mathematics, modern mathematicians such as Cantor, Hausdorff, Julia, Koch, Peano, and Sierpinski extended the concept of self-similarity to include objects that were self-similar on an infinite number of scales. Objects like those in Figure 5.1.3 are often used to suggest the nature of such objects, even though the objects as shown are only self-similar on a few levels. Illustrations such as these are used because no technology is capable of creating a figure revealing self-similarity on an infinite number of scales. (Why?)

URLs
5.1.1
–
5.1.6

Figure 5.1.3 Self-similarity on Many Scales

URL
5.1.7

For the first half of the twentieth century, mathematical objects with this property were generally treated as curiosities or "mathematical monsters" with few if any significant applications. One of the first mathematicians to recognize in these curiosities the seeds of a new mathematical discipline was Benoit Mandelbrot. Mandelbrot coined the term "fractal" for objects that are self-similar at all scales. In his landmark book *The Fractal Geometry of Nature* published in 1982 by W. H. Freeman and Company, Mandelbrot defined the term fractal as follows.

Definition 5.1.2 A *fractal* is a rough or fragmented geometric shape that can be subdivided in parts, each of which is (at least approximately) a smaller version of the whole.

Figure 5.1.4 Mandelbrot Set

One of the most famous fractals now bears Mandelbrot's name, the Mandelbrot set (see Figure 5.1.4). Mandelbrot's discoveries helped focus the attention of mathematicians, scientists, engineers, computer scientists, and many other professionals on fundamental questions related to fractal geometry and chaotic systems in general. As a result, many important discoveries were made and applications developed that have changed the way we think about mathematics.

Why is geometry often described as "cold" and "dry?" One reason lies in the inability to describe the shape of a cloud, a mountain, a coastline or a tree. Clouds are not spheres, coastlines are not circles, and bark is not smooth, nor does lightning travel in a straight line.... The existence of these patterns challenges us to study those forms that Euclid leaves aside as being "formless," to investigate the morphology of the "amorphous."

—*Benoit Mandelbrot*

Over the last decade, physicists, biologists, astronomers and economists have created a new way of understanding the growth and complexity of Nature. This new science, called chaos, offers a way of seeing order and pattern where formerly only random, erratic, the unpredictable—in short, the chaotic—had been observed.

—*James Gleick*

Fractal geometry will make you see everything differently. There is a danger in reading further. You risk the loss of your childhood vision of clouds, forests, flowers, galaxies, leaves, feathers, rocks, mountains, torrents of water, carpets, bricks, and much else besides. Never again will your interpretation of these things be quite the same.

—*Michael F. Barnsley*

> Now, as Mandelbrot points out… Nature has played a great joke
> on the mathematicians. The same pathological structures the math-
> ematicians invented to break loose from 19th-century naturalism
> turn out to be inherent in familiar objects all around us in Nature.
> —*Freeman Dyson*

Today, fractal geometry is used in a variety of applications. For
instance, in the sciences and engineering fractal geometry is used to characterize the shape and texture of complex surfaces. In film and television,
it is used to create realistic imaginary landscapes that serve as the back-
drop for science fiction movies. For example, Figure 5.1.5 shows an artifi-
cial landscape created using fractal geometry.

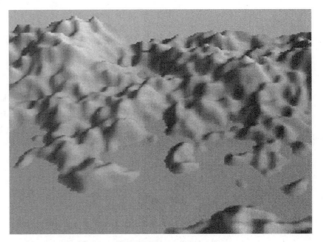

Figure 5.1.5 Fractal Landscape

5.1 Introduction to Self-similarity

In this section, you will . . .

- **Investigate the concept of self-similarity using graphical, numerical, and algebraic approaches and representations.**

The Koch Snowflake Curve

URL
5.1.8

Figure 5.1.6 shows the first three steps in the development of the Koch snowflake curve. The process begins with a segment connecting two points ($n = 0$). In the first iteration of the process, the original segment is replaced by a curve called the *generator* of the Koch snowflake curve ($n = 1$). In the second iteration of the process, each segment of the generator ($n = 2$) is replaced by a smaller version of itself ($n = 3$). By repeating this process an infinite number of times, a true fractal is obtained.

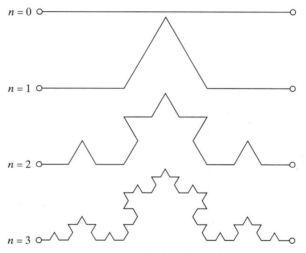

Figure 5.1.6 Koch Snowflake Curve

Definition
5.1.3

The *generator* of a given fractal is a curve that defines the transformation from one iteration. At each iteration, every segment of the developing fractal curve is replaced with a suitably scaled version of the generator.

A more interesting version of this process is shown in Figure 5.1.7, where the generator is initially applied to the sides of an equilateral triangle. The graphics were created using Logo, a computer language developed for elementary and middle school children. The Logo programming language is available from a number of sources. *MSW Logo*, included on the *Active Geometry* CD-ROM, is a convenient tool for creating and exploring fractals. For instance, the Logo code in Table 5.1.1 captures the iterative nature of this process. The commands flake 1, flake 2, flake 3, and flake 4 were used to generate the curves in Figure 5.1.7.

**URL
5.1.9**

**CD
5.1.1**

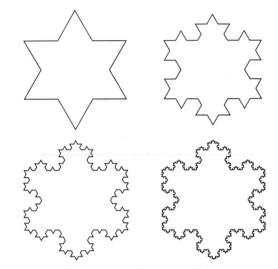

Figure 5.1.7 Koch Snowflake

Logo Code	Comments

to flake :n
  pu ht rt 60 bk 90 lt 30 pd
  make "x 1
  repeat :n [make "x 3*:x]
  make "l 200/:x
  repeat 3[ifelse :n = 0 [fd :l]
  [line :n :l] rt 120]
  pu home pd setfc :n fill
end
to line :n :l
  ifelse :n = 1 [fd :l lt 60 fd :l
  rt 120 fd :l lt 60 fd :l] [line
  :n-1 :l lt 60 line :n-1 :l rt
  120 line :n-1 :l lt 60 line :n-
  1 :l]
end
	The name of the main procedure is **to flake :n.** The notation **:n** indicates a variable requiring a keyboard input, entered as flake **1,** flake **2,** flake **3,** and so on.
	
	The length of the original segment is set by **make "l 200/:x**
	
	The original triangle is drawn by **repeat 3[ifelse :n = 0 [fd :l] [line :n :l] rt 120]** Do *not* insert a carriage return in this line.
	
	The body of the procedure **to line :n :l** is a single line. Do *not* insert carriage returns at the end of the first, second, or third lines.

Table 5.1.1 Logo Code for Koch Snowflake Curve

Figure 5.1.8 uses shading to highlight successive additions to the Koch snowflake from one level of iteration to another. This figure also motivates an investigation of the perimeter and areas of successive iterates of the process.

Figure 5.1.8 Shaded Koch Snowflake

Perimeter and Area of the Koch Snowflake Curve

Every iteration of the process used to construct the Koch snowflake curve replaces each segment with a version of the generator $4/3$ as long as the replaced segment. (Why?) Consequently, the length of the curve increases by a factor of $4/3$ with each iteration. Table 5.1.2 summarizes this process for the Koch snowflake. The length of each side of the original equilateral triangle is represented by the variable L. As a result, the perimeter of the original triangle is $3L$ and the perimeter of the n-th iteration Koch curve is $3L(4/3)^n$. Since $\lim_{d \to \infty} r^d = \infty$ as $n \to \infty$ for $r > 1$, the perimeter of the Koch snowflake increases without bound as n goes to infinity. So, the perimeter of the Koch snowflake is infinite and could therefore never be drawn or traced by any means.

N	# Seg.	Seg. Length	Perimeter
0	$4^0(3) = 3$	$L = L/3^0$	$3L$
1	$4^1(3) = 12$	$L/3 = L/3^1$	$12(L/3) = 4L$
2	$4^2(3) = 48$	$L/9 = L/3^2$	$48(L/9) = (16/3)L$
3	$4^3(3) = 192$	$L/27 = L/3^3$	$192(L/27) = (48/9)L$
n	$4^n(3)$	$L/3^n$	$4^n(3)L/3^n = 3L(4/3)^n$

Table 5.1.2 Perimeter of the Koch Snowflake

As suggested by Figure 5.1.8, successive iterations of the process add smaller and smaller triangular regions to the snowflake. Table 5.1.3 shows the progression of added areas.

If the area of the original equilateral triangle is assumed to be one square unit, the additional area added on the n-th iteration is given by

N	# Seg.	Triangles Added	Additional Area
0	3		
1	12	3	$3(1/9)$
2	48	12	$12(1/9)^2$
3	192	48	$48(1/9)^3$
n	$4^n(3)$	$4^{n-1}(3)$	$4^{n-1}(3)(1/9)^n$

Table 5.1.3 Area of the Koch Snowflake

$4^{n-1}(3)(1/9)^n$. The sum of the areas of all such regions plus the area of the original equilateral triangle is given by the expression

$$1 + \sum_{i=1}^{n} 4^{i-1}(3)\left(\frac{1}{9}\right)^i = 1 + \frac{1}{3} \sum_{i=1}^{n} \left(\frac{4}{9}\right)^{i-1}$$

Taking the limit of this expression as n goes to infinity yields

$$\lim_{n \to \infty} 1 + \frac{1}{3} \sum_{i=1}^{n} \left(\frac{4}{9}\right)^{i-1} = 1.6$$

So, while the perimeter of the Koch snowflake fractal is infinite, its area is finite. This result is both startling and intriguing.

CD 5.1.2   In Tables 5.1.4 and 5.1.5, the formulas for perimeter and area of the snowflake curve seen in Tables 5.1.2 and 5.1.3 are implemented in an *MS Excel* spreadsheet. Using *Excel's* Fill Down option, this table may be extended through hundreds of iterations.

N	# Seg.	Seg. Length	Perimeter	Added Area	Total Area
0	3	1.000000000	3.000000000	0.000000000	1.000000000
1	12	0.333333333	4.000000000	0.333333333	1.333333333
2	48	0.111111111	5.333333333	0.148148148	1.481481481
3	192	0.037037037	7.111111111	0.065843621	1.547325103
4	768	0.012345679	9.481481481	0.029263832	1.576588935
5	3072	0.004115226	12.64197531	0.013006147	1.589595082
6	12288	0.001371742	16.85596708	0.005780510	1.595375592

Table 5.1.4 Perimeter and Area of the Koch Snowflake

N	Segments	Length	Perimeter	Added Area	Total Area
0	= 3*4^A2	= 1/3^A2	= 3*(4/3)^A2	0	1
= A2+1	= 3*4^A3	= 1/3^A3	= 3*(4/3)^A3	= 3*(1/9)^A3*4^(A3−1)	= F3+G2

Table 5.1.5 Formulas for Table 5.1.4

Additional Examples of Fractal Curves

Example 5.1.1

The Koch anti-snowflake curve (see Figure 5.1.9) is created by reversing the orientation of the generator used in the Koch snowflake, causing the fractal to grow inward rather than outward. The Logo code for the curve is shown in Table 5.1.6. Finding expressions for the perimeter and area of this fractal is left as an exercise.

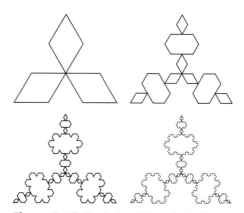

Figure 5.1.9 Koch Anti-Snowflake Curve

<table>
<tr><td>Logo Code</td></tr>
<tr><td>Note: The body of the procedure to line :n :l is a single line. Do not insert carriage returns at the end of the first or second line of the code shown below.</td></tr>
</table>

```
to flake :n
    pu ht rt 60 bk 90 lt 30 pd
    make "x 1
    repeat :n [make "x 3*:x]
    make "l 200/:x
    repeat 3 [ifelse :n=0 [fd :l] [line :n :l] rt 120]
    pu home pd setfc :n fill
end
to line :n :l
    ifelse :n=1 [fd :l rt 60 fd :l lt 120 fd :l rt 60 fd :l ]
    [line :n-1 :l rt 60 line :n-1 :l lt 120 line :n-1 :l rt 60
    line :n-1 :l]
end
```

Table 5.1.6 Logo Code for Koch Anti-Snowflake Curve

CD
5.1.3

In Tables 5.1.7 and 5.1.8, the formulas for perimeter and area of the anti-snowflake curve are implemented in an *MS Excel* spreadsheet.

N	# Seg.	Seg. Length	Perimeter	Added Area	Total Area
0	3	1.000000000	3.000000000	0.000000000	1.000000000
1	12	0.333333333	4.000000000	0.333333333	0.666666667
2	48	0.111111111	5.333333333	0.148148148	0.518518519
3	192	0.037037037	7.111111111	0.065843621	0.452674897
4	768	0.012345679	9.481481481	0.029263832	0.423411065
5	3072	0.004115226	12.64197531	0.013006147	0.410404918
6	12288	0.001371742	16.85596708	0.005780510	0.404624408
7	49152	0.000457247	22.47462277	0.002569116	0.402055292
8	196608	0.000152416	29.96616369	0.001141829	0.400913463

Table 5.1.7 Perimeter and Area of the Anti-Snowflake

N	# Seg.	Seg. Length	Perimeter	Added Area	Total Area
0	=3*4^A2	=1/3^A2	=3*(4/3)^A2	0	1
= A2+1	=3*4^A3	=1/3^A3	=3*(4/3)^A3	=3*(1/9)^A3*4^(A3−1)	=G2−F3

Table 5.1.8 Formulas for Table 5.1.8

Example
5.1.2

The fractal shown in Figure 5.1.10 is based on a hat-shaped generator applied to the sides of a square. Logo code for the fractal is shown in

Figure 5.1.10 Hat Generator

Table 5.1.9. Does this curve ever cross itself? Finding an expression for the perimeter and area of this fractal is left as an exercise.

Logo Code

Note: The body of the procedure to line2 :n :l is a single line. Do *not* insert carriage returns at the end of the first or second line of the code shown below.

```
to flake2 :n
    pu hr t rt 60 bk 120 lt 60 pd
    make "x 1
    repeat :n [make "x 3*:x]
    make "l 150/:x
    repeat 4 [ifelse :n=0 [fd :l] [line2 :n :l] rt 90]
    pu home pd setfc
end
to line2 :n :l
    ifelse :n=1 [fd :l lt 90 fd :l rt 90 fd :l rt 90 fd :l lt 90 fd :l]
    [line2 :n-1 :l lt 90 line2 :n-1 :l rt 90 line2 :n-1 :l rt 90 line2 n-1 :l
    lt 90 line2 :n=1 :l]
end
```

Table 5.1.9 Logo Code for Hat Curve

Example 5.1.3

URL 5.1.10

The fractal shown in Figure 5.1.11 is known as the Peano plane filling curve. As the process producing this curve is iterated over and over, a shape resembling a square begins to emerge. Further iterations trace the curve through more and more interior points of that square. As $n \to \infty$, every point in the square becomes a limit point of the Peano curve. Since no point in the square is "missed" by this process, the curve is called "plane filling." Finding an expression for the length of the curve and a Logo code to draw it are left as an exercise.

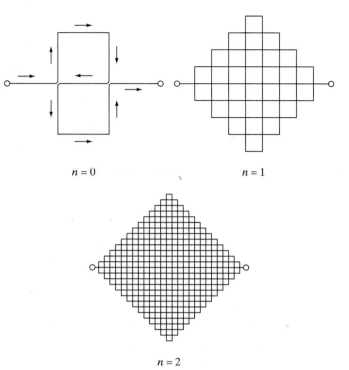

$n = 0$ $n = 1$

$n = 2$

Figure 5.1.11 Peano Curve

Summary

Fractals such as the Koch snowflake have features that are clearly non-Euclidean in nature: Finite areas contained within infinite perimeters; self-similarity on an infinite number of scales; infinitesimal segments (why?); and more. While graphical tools and spreadsheets are powerful tools in the development of mathematical concepts and insights, they can never create a true fractal or model its features comprehensively. Nor can we envision a true fractal in all its detail. But using the power of mathematics, we may investigate the structure of such objects and understand their most significant features.

Definitions	
5.1.1	Any object composed of smaller models of itself is said to be *self-similar.*
5.1.2	A *fractal* is a rough or fragmented geometric shape that can be subdivided in parts, each of which is (at least approximately) a smaller version of the whole.
5.1.3	The *generator* of a given fractal is a curve that defines the transformation from one iteration. At each iteration, every segment of the developing fractal curve is replaced with a suitably scaled version of the generator.

Table 5.1.10 Summary, Section 5.1

URLs		Note: Begin each URL with the prefix http://
5.1.1	h	www-history.mcs.st-and.ac.uk/history/Mathematicians/Cantor.html
5.1.2	h	www-history.mcs.st-and.ac.uk/history/Mathematicians/Hausdorff.html
5.1.3	h	www-history.mcs.st-and.ac.uk/history/Mathematicians/Julia.html
5.1.4	h	www-history.mcs.st-and.ac.uk/history/Mathematicians/Koch.html
5.1.5	h	www-history.mcs.st-and.ac.uk/history/Mathematicians/Peano.html
5.1.6	h	www-history.mcs.st-and.ac.uk/history/Mathematicians/Sierpinski.html
5.1.7	h	www-history.mcs.st-and.ac.uk/history/Mathematicians/Mandelbrot.html
5.1.8	cs	http://cut-the-knot.com/do_you_know/dimension.html#sdim
5.1.9	s	http://www.softronix.com/
5.1.10	cs	cut-the-knot.com/do_you_know/hilbert.html

Table 5.1.11 Section 5.1 URLs (c = concept, h = history, s = software, d = data)

Exercises

1. Using each of the following curves as a generator, sketch the next level of iteration.

   a) 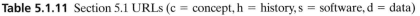 b) , where the angle formed is 120°.
2. Find an expression for the length of the curve in the *n*th iteration of the fractal in
   a) Example 5.1.1
   b) Example 5.1.2
   c) Example 5.1.3
3. Estimate the area of the squares formed in the first three iterations of the fractal in Example 5.1.3 given that the original end points are three units apart.
4. Using *MSW Logo*, create drawings of the 2nd, 3rd, and 4th iterations of the generators in exercise 1.
5. Estimate the curve lengths of the first five iterations of generator b) in exercise 1, assuming that the generators are composed of segments of equal length.
6. Find examples of vegetables and other natural objects with self-similar properties.

Investigations

Koch Snowflake Curve 1
Tool(s) Geometers Sketchpad   *Data File(s)* cd5_1_4.gsp
                                                          cd5_1_4a.gss

Focus
The Koch snowflake curve is generated using an iterative process in which the same geometrical transformation is applied over and over on smaller and smaller scales.

Tasks
1. Under the File pull-down menu, select New Sketch, then Open the file cd5_1_4.gsp. In the New Sketch window, select two points. Switching to the cd5_1_4a.gss window, click on the Fast button. A Depth of Recursion dialogue box will open. Generate a sequence of curves by entering the numbers 0, 1, and 2. For each case, determine the number of segments.
2. If this process could be continued an infinite number of times, what would happen to the number of segments? To the length of the segments? To the length of the curve?

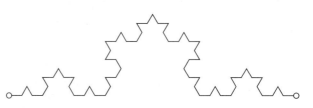

Koch Snowflake Curve 2
Tool(s) Geometers Sketchpad   *Data File(s)* cd5_1_4.gss

Focus
The Koch snowflake curve is self-similar, that is, small pieces resemble larger pieces.

Tasks
1. Using the point at the left end of the $n = 0$ curve, drag the curve down onto the $n = 1$ curve. How many triangles are added to the $n = 0$ curve to obtain the $n = 1$ curve?
2. Using the point at the left end of the $n = 1$ curve, drag the curve down onto the $n = 2$ curve. How many triangles are added to the $n = 1$ curve to obtain the $n = 2$ curve?
3. How many triangles would be added to the $n = 2$ curve to obtain the $n = 3$ curve? Test your answer by using cd5_1_4a to draw the $n = 3$ curve between the endpoints of the $n = 1$ curve, then dragging the $n = 3$ curve down onto to $n = 2$ curve.

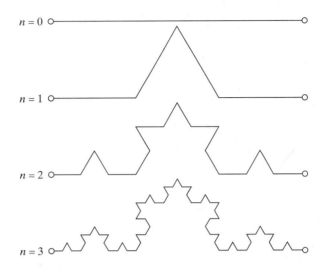

Koch Snowflake Curve 3
Tool(s) Geometers Sketchpad   *Data File(s)* cd5_1_6.gsp

Focus
As *n* goes to infinity, what happens to the Koch snowflake curve length and area?

Tasks
1. Compute segment and curve lengths for $n = 0$, 1, and 2. Find an expression for the length of the segments in the *n*th iteration. For the length of the curve in the *n*th iteration.
2. Compare shaded area for $n = 0$, 1, and 2. Find an expression for the shaded area in the *n*th iteration.
3. As *n* goes to infinity, what happens to the curve length? To the shaded area?

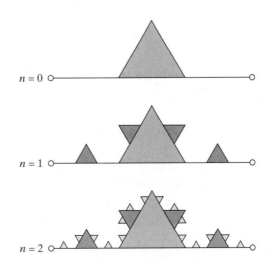

Sierpinski Triangle 1
Tool(s) Geometers Sketchpad   *Data File(s)* cd5_1_7.gsp
                                              cd5_1_7a.gss

Focus
The Sierpinski triangle is generated using an iterative process in which the same geometrical transformation is applied over and over on smaller and smaller scales.

Tasks
1. Under the File pull-down menu, select New Sketch, then Open the file cd5_1_7.gsp. In the New Sketch window, select three points. Switching to the cd5_1_7a.gss window, click on the Fast button. A Depth of Recursion dialogue box will open. Generate a sequence of triangles by entering the numbers 0, 1, and 2. For each case, determine the number of shaded triangles.
2. If this process could be continued an infinite number of times, what do you think would happen to the number of triangles? To the perimeter of the triangles? To the area of the triangles?

Sierpinski Triangle 2
Tool(s) Geometers Sketchpad   *Data File(s)* cd5_1_8.gsp

Focus
As n goes to infinity, what happens to the Sierpinski triangles' perimeters and areas?

Tasks
1. Compare the areas of each shaded triangle for $n = 0, 1$, and 2. Find an expression for the sum of the areas of the shaded triangles in the nth iteration.
2. Compare the perimeters of each shaded triangle for $n = 0, 1$, and 2. Find an expression for the sum of the perimeters of the shaded triangles in the nth iteration.

3. As *n* goes to infinity, what happens to the sum of the areas of the shaded triangles? To the sum of the perimeters of the shaded triangles?

Size = 1.31 Inches

5.2   Fractal Dimension

In this section,
you will . . .

- **Investigate the concept of fractal dimension;**
- **Compute the fractal dimension of curves using the ruler method;**
- **Compute the fractal dimension of curves using the grid method;**
- **Compute the fractal dimension of curves based on measurements of their generators.**

The Trouble with Coastlines

One of the questions that motivated Mandelbrot to develop new ways of thinking about complex curves was, "How long is the coastline of Britain?" As suggested by the sample coastline in Figure 5.2.1, coastlines may be complex and difficult to measure.

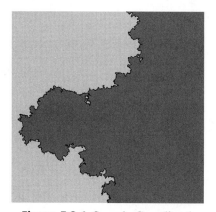

Figure 5.2.1 Sample Coastline 1

One of the issues associated with measuring curves of this sort is the length of the "ruler" used. For instance, Figure 5.2.2 shows two copies of the same coastline, each measured with a ruler of a different length. One of the rulers is 50 units in length, the other 20. Notice that both rulers cut across bays and promontories.

It is natural to assume that by choosing a short enough ruler, this sort of problem may be avoided and an accurate measurement obtained. For instance, a ruler that is one meter in length might cut across rocks and puddles, but will not cut across bays and promontories. Of course, the coastline is not a well-defined feature of the landscape. Depending on tides and other dynamic coastal processes, coastal features may change on any given day on the scale of a meter or so. On the other hand, it would be a rare day for any coastline to change on the scale of a bay or promontory. So, while the question, "How long is the coastline of Britain?" may have a reasonable or useful answer, it cannot have a mathematically precise answer.

The question becomes much more interesting when the curves being measured are fractals. Because fractals are self-similar on every scale, zooming in reveals meaningful "bays and promontories" on every scale.

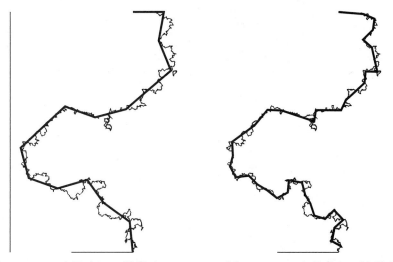

Measurement 1: Ruler = 50 Units   Measurement 1: Ruler = 20 Units
Figure 5.2.2 Sample Measurements

In coastlines, Mandelbrot found a metaphor for such curves years before he first used the term "fractal." In attempting to characterize the complexity of such coastlines, he devised the concept of "fractal dimension."

Euclidean Dimensionality

In Figure 5.2.3, a unit line segment, square, and cube are divided into subunits, each having dimensions one-third the length of the segment. Table 5.2.1 relates the topological dimension (D) of each object, the length of the unit of measurement ($1/L$), and the number of linear, square, or cubic subunits (N) in each object. Data is also included for objects that are based on segments divided into four and five subunits. In general, the relationship between N, L, and D is given by $N = [1/L]^D$. Taking the logarithm of both sides of this equation and solving for D yields the equation $D = \log N/\log(1/L)$.

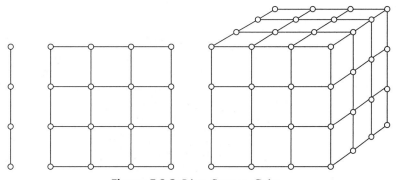

Figure 5.2.3 Line, Square, Cube

Object	L	D	$(1/L)^D = N$
Segment	1/3	1	$[1/(1/3)]^1 = 3$
	1/4		$[1/(1/4)]^1 = 4$
	1/5		$[1/(1/5)]^1 = 5$
Square	1/3	2	$[1/(1/3)]^2 = 9$
	1/4		$[1/(1/4)]^2 = 16$
	1/5		$[1/(1/5)]^2 = 25$
Cube	1/3	3	$[1/(1/3)]^3 = 27$
	1/4		$[1/(1/4)]^3 = 64$
	1/5		$[1/(1/5)]^3 = 125$

Table 5.2.1 Euclidean Dimension

Example 5.2.1 The sides of a cube are divided into 10 linear subunits. How many cubic subunits are there? Using $N = [1/L]^D$, $N = [1/(1/10)]^3 = 1000$ cubic subunits.

Fractal Dimension: Grid Method

In Euclidean geometry, curves consisting of a finite number of line segments are 1-dimensional. Does it seem appropriate to extend the Euclidean concept of dimensionality to fractal curves? Would you describe Koch's curve as 1-dimensional? What about Peano's plane-filling curve (see Figure 5.2.4)? Iterated an infinite number of times, Peano's curve fills a 2-dimensional square. Does it seem more reasonable to think of Peano's curve as being 1-dimensional or 2-dimensional?

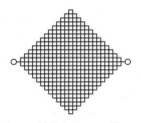

Figure 5.2.4 Peano Curve

In general, trying to extend Euclidean concepts of dimensionality to describe fractal objects forces one to choose between unacceptable alternative characterizations. This circumstance illustrates the need for an alternative concept of dimensionality better suited to characterizing the complexity of fractal objects. Because the distinguishing characteristic of fractals is self-similarity on an infinite number of scales, zooming in on a given fractal reveals essentially the same structure at any scale. What varies from fractal to fractal is the structure of the fractal relative to the scale of observation; that is, some fractals are "busier" than other fractals. For instance, Figure 5.2.5 shows three coastline curves, each

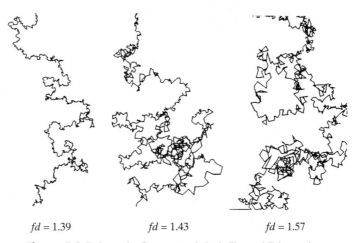

$fd = 1.39$ $fd = 1.43$ $fd = 1.57$

Figure 5.2.5 Sample Curves and their Fractal Dimensions

composed of 1-dimensional segments arranged in a complex pattern. The more complex the pattern, the greater the numerical value of the fractal dimension(fd). Fractal dimension provides more than an ordinal basis of comparison between curves, such as "this curve is more complex than that curve." Differences in the fractal dimensions represent quantifiable differences in complexity.

There are several methods for investigating the complexity of a curve. One method begins by overlaying the object with a sequence of grids of different sizes (see Figures 5.2.6 and 5.2.7 and cd5_2_1). For each grid, the cells intersected by the fractal are counted, providing an index of the relative complexity of the object on different scales. If one could apply this procedure to a true fractal, using finer and finer grids would reveal greater and greater structural detail. Furthermore, a non-linear relationship would emerge between the length $(1/L)$ of the sides of the grid cells and the number (N) of cells intercepted by the curve (see Table 5.2.2) in which the sequence of values given by the ratio $\log N/\log(1/L)$ converges to a fixed value, the absolute value of which is called the *fractal dimension* of the object.

**CD
5.2.1**

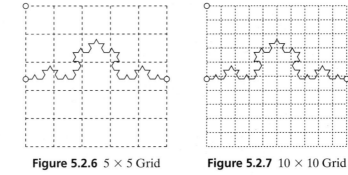

Figure 5.2.6 5×5 Grid **Figure 5.2.7** 10×10 Grid

L	**1/L**	**N**	**log(N)/log(1/L)**
5	0.2	7	−1.209061955
10	0.1	24	−1.380211242
20	0.05	46	−1.278031896
40	0.025	89	−1.216802128

Table 5.2.2 Grid Data: Koch Curve

Definition 5.2.1

The *fractal dimension* of a self-similar object is given by the absolute value of the ratio $\log N/\log(1/L)$, where $(1/L)$ is the size of the grid cell or ruler used to measure the object and N is the number of grid cells or rulers used in measuring the object.

Because it is impossible to draw a true fractal curve, any application of this procedure ultimately leads to a cell size that is smaller than the segment lengths in the particular representation of the self-similar curve being tested. At that point, the process ceases to reveal additional details of the curve's structure. In practice, this sets a lower limit on useable grid sizes for a given fractal. In Table 5.2.2, that limit is exceeded at the value $1/L = 0.025$ units. Because it is not always obvious at which scale this breakdown occurs, data is normally gathered on several scales and analyzed graphically to obtain an estimate of the fractal dimension of the object.

URL 5.2.1

A wonderful teaching tool for exploring this process is the Java Applet *Fractal Coastlines* created by the Patterns in Nature Project at the Center for Polymer Studies, Boston University. The grid method for determining fractal dimension is based on a simple concept: Repeated measurements on smaller and smaller scales to gather information on finer and finer details. This method is based on concepts and procedures first published in 1919 by Hausdorff in connection with his theory of topological and metric spaces.

Figure 5.2.8 shows a sample coastline to be analyzed using the grid method. In Figures 5.2.9 and 5.2.10 (on page 224), grids of two different sizes are superimposed over the coastline. Each grid square containing one or more points of the coastline is marked by shading then counted. Figure 5.2.11 (on page 224) contains a table of data gathered using this approach. When data of this sort is plotted and a curve fitted to the data, equations of the form $N = [1/L]^D$ are obtained in which D takes on values that are not integers. The absolute value of the exponent in this equation is interpreted as the fractal dimension, in this case 1.22.

Fractal Dimension: Ruler Method

In considering the dimensionality of coastlines, Mandelbrot was interested in what might be learned from available data. Consequently, he took repeated measurements on coastline maps using rulers of different lengths

Figure 5.2.8 Sample Coastline 2

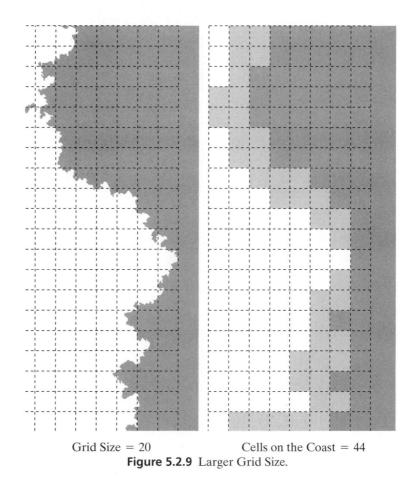

Grid Size = 20   Cells on the Coast = 44
Figure 5.2.9 Larger Grid Size.

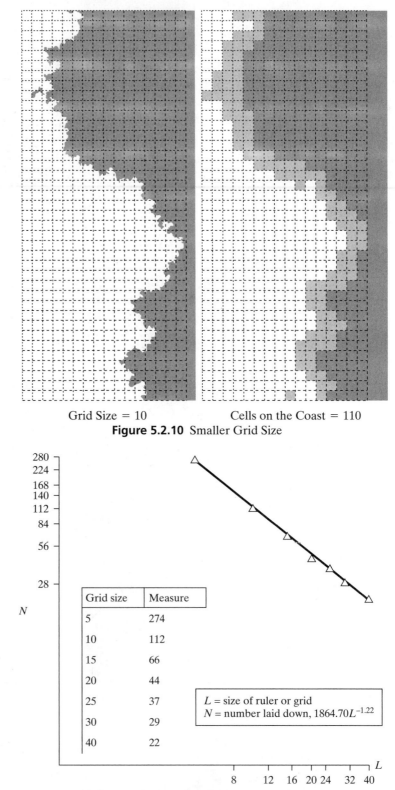

Grid Size = 10   Cells on the Coast = 110
Figure 5.2.10 Smaller Grid Size

Grid size	Measure
5	274
10	112
15	66
20	44
25	37
30	29
40	22

L = size of ruler or grid
N = number laid down, $1864.70L^{-1.22}$

Figure 5.2.11 Fractal Dimension: Grid Method

(see Figure 5.2.2). When data of this sort is plotted and a curve fitted to the data, the same sort of equations are obtained as when using the grid method. For instance, in the case of the data plotted in Figure 5.2.12, the equation $N = 2014.80L^{-1.27}$ is obtained.

An Analytic Approach to Fractal Dimension

If the ruler or grid method is used to determine the fractal dimension of the Koch snowflake curve, the value obtained is approximately the same as that obtained for the coastline examples. One interpretation of this finding is that all three curves exhibit comparable sorts of complexity (see Figure 5.2.13). What do you think?

Because the Koch curve is a true fractal, repeating the same geometric "theme" on an infinite number of scales, an analytic approach may be used to determine its fractal dimension. Figure 5.2.14 shows the generator for the Koch snowflake curve. In terms of the notations used in Table 5.2.3, the fractal dimension of the Koch snowflake curve may be computed as $D = \log N/\log(1/L) = \log 4/\log 3 \approx 1.26$. $N = 4$ because there are four subunits to the curve and $(1/L) = 3$ because each subunit L is one-third the distance between the beginning and end points of the curve.

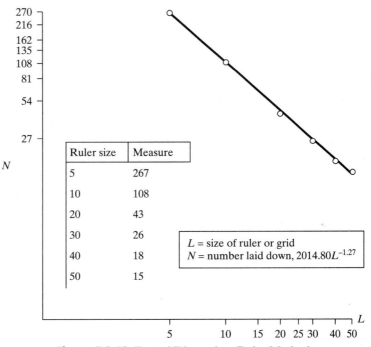

Ruler size	Measure
5	267
10	108
20	43
30	26
40	18
50	15

L = size of ruler or grid
N = number laid down, $2014.80L^{-1.27}$

Figure 5.2.12 Fractal Dimension: Ruler Method

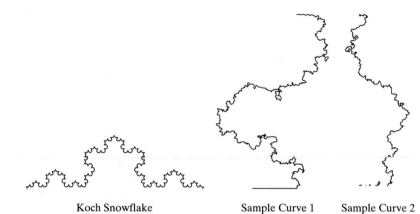

Koch Snowflake   Sample Curve 1   Sample Curve 2

Figure 5.2.13 Comparable Complexity

Figure 5.2.14 Koch Generator

Example 5.2.2 Find the fractal dimension of the Koch anti-snowflake curve.

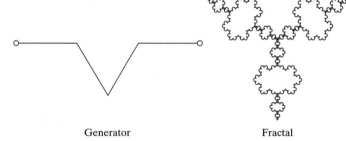

Generator   Fractal

$$\log 4/\log 3 \approx 1.26$$

Example 5.2.3 Find the fractal dimension of the hat curve.

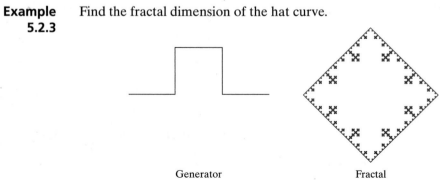

Generator   Fractal

$$\log 5/\log 3 \approx 1.46$$

**Example
5.2.4**   Find the fractal dimension of Peano's curve.

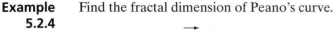

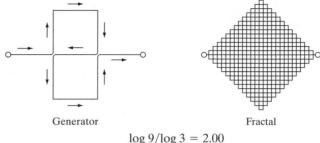

Generator   Fractal

$$\log 9/\log 3 = 2.00$$

Summary

Fractal dimension is a relatively new mathematical concept. A variety of computational techniques have been developed to estimate the fractal dimension of curves and surfaces. Of these approaches, the simplest conceptual models are the ruler and grid methods outlined in the section. Each method seeks to characterize the complexity of fractal objects by gathering information to the extent to which the objects utilize available space at different scales. In general, fractal curves have fractal dimensions between 1 and 2. As one might expect, fractal surfaces have fractal dimensions between 2 and 3. By quantifying the degree of complexity exhibited by a given fractal, fractal dimension makes it easier to talk about the relative complexity of two or more curves or surfaces.

Definitions	
5.2.1	The fractal dimension of a self-similar object is by the absolute value of the ratio $\log N/\log(1/L)$ where $(1/L)$ is the size of the grid cell or ruler and N is the number of grid cells or rulers used in measuring the object.

Table 5.2.3 Summary, Section 5.2

URL	Note: Begin each URL with the prefix http://
5.2.1 s	polymer.bu.edu/java/java/coastline/coastline.html

Table 5.2.4 Section 5.2 URLs (c = concept, h = history, s = software, d = data)

Exercises

1. Sketch the next two iterations of each of the following generators. Create your own generator and repeat the process.

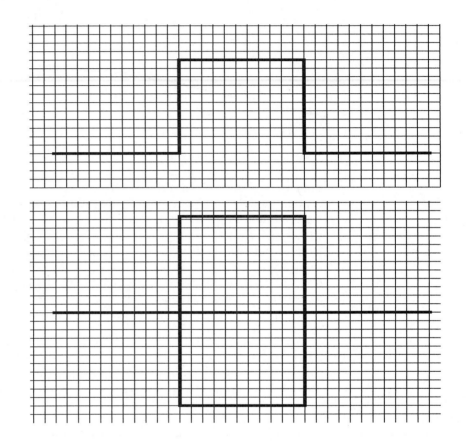

2. Sketch the next two iterations of the following generator. Create your own generator and repeat the process.

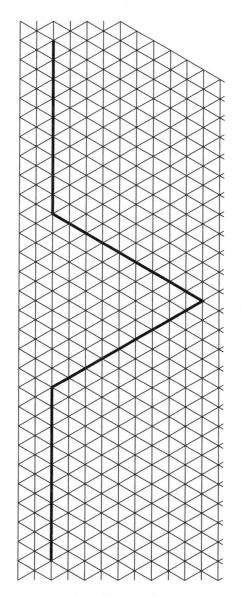

Investigations

Using Fractal Coastline
Tool(s) Fractal Coastline   *Data File(s) none*

Focus
"How long is the coastline of England?" The question that motivated Mandelbrot to formulate the concept of fractal dimension may be investigated using the Fractal Coastline applet at http://polymer.bu.edu/java/ java/coastline/coastline.html

Tasks
1. Start *Fractal Coastline* and set the sliders at 10 iterations, 2 starting points, and a roughness of 0.20. Every time you select Draw Coastline, you will obtain a new coastline. Do so until you obtain a coastline that seems interesting.
2. You will use two methods to determine the fractal dimension of your coastline. The first method measures the length of the coastline using rulers of different length. The second method covers the coastline using grids of different sizes. Select Ruler in the Measure pull-down menu.
3. A Ruler Control window will appear. Set the ruler length to 50 units and Auto Measure the coastline. When the Ruler Control window reappears, decrease the size of the ruler to 45 units and Auto Measure again. Repeat this procedure over and over until you have collected at least 8 measurements. Click Done Measuring
4. A window will appear with the options Measure, Table, and Graph. Select Table to see the data gathered. Select Graph to see a plot of the data. Select Ruler Data and Linear Graph (log-log scale). Finally select Curve Fit. An equation will appear in which the absolute value of the exponent is interpreted as the fractal dimension of the coastline.
5. Select Coastline and repeat the entire process using the grid method.
6. Compare the two values obtained for the fractal dimension of the coastline. Which do you believe to be a better method? Why?

Determining Fractal Dimension by Counting Grids
Tool(s) Geometers Sketchpad   *Data File(s) cd5_2_1.gsp*

Focus
Using the grid method (box counting by hand) to determine the fractal dimension of the Koch snowflake.

Tasks
1. Count the cells intersected by the Koch snowflake curve in each of the grids shown (see cd5_2_1.gsp). Enter the data into a table.
2. Using *MS Excel* or a comparable spreadsheet, find the log of each entry in the data table.

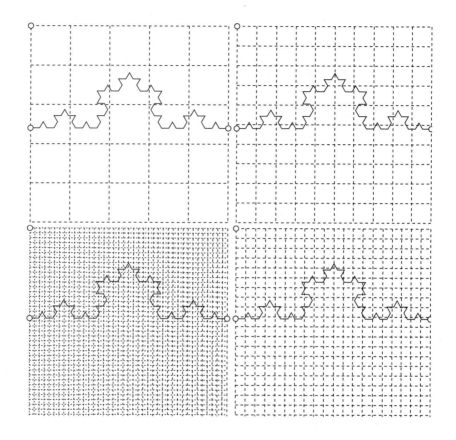

3. Perform a linear regression on the transformed data to obtain the curve of best fit. Use the box size data as the *x*-variable and the box count data as the *y*-variable. In the regression equation, the absolute value of the coefficient of *x* is the fractal dimension.

Determining Fractal Dimension by Java Applet
Tool(s) Fractal Dimension Macros   *Data File(s) none*

Focus
The fractal dimension of the Koch snowflake may be investigated using Matthew Warfel's fractal dimension applet at
http://www.cee.cornell.edu/~mdw/fractech.html

Tasks
1. Select Box Counting from the frame at the left side of the screen.
2. Select Fast from the frame at the left side of the screen.
3. Which do you believe to be the better method? Why?

Boundary Fractal Dimension of Koch Snowflake by Box-Counting
Method

Calculated D	Theoretical D	% error
1.2866	1.262	2.5

Box Size	Box Count
4	233
6	146
8	105
10	80
12	62
14	48
16	46
18	34

Working With Real Coastlines
Tool(s) USGS Coastline Extractor   *Data File(s) none*

Focus
The United States Geological Survey (USGS) maintains on-line files of
the entire U.S. Coastline at http://crusty.er.usgs.gov/coast/getcoast.html.
Using hand measurement techniques, the fractal dimension of real coastlines may be explored.

Tasks
1. Print out a copy of the map of Boston Harbor shown below from the
   USGS site. Overlay the figure with several grids of different sizes.
   Using the procedures in Investigation 2 above, estimate the fractal
   dimension of the coastline.
2. Using the USGS on-line information, select a different section of the
   U.S. coast and estimate its fractal dimension.

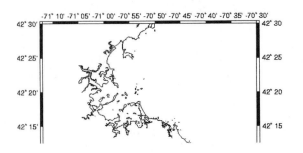

5.3   Iterated Function Systems

In this section,
you will . . .

- **Investigate the concept of an iterated function system;**
- **Create fractals using iterated function systems.**

Sections 5.1 and 5.2 introduced the Koch snowflake and other fractal curves using the *Geometers Sketchpad* and the Logo programming language. Using only these technologies, one might conclude that the only way to approximate such curves is by plotting carefully sequenced line segments. In fact, a far more general approach exists for creating fractals based on iterated function systems (IFS). This section focuses on systems of linear transformations operating in the Euclidean plane. The following definitions are useful in discussing iterated function systems.

Definition
5.3.1

A *space S* is a set. The elements of the set are the *points* of the space.

Definition
5.3.2

A *metric space* (S, d) consists of a space S and a function d that associates a real number with any two elements of S. The function d must have the following properties:

$$d(x, y) = d(y, x) \quad \forall \quad x, y \in S$$

$$0 < d(x, y) < \infty \quad \forall \quad x, y \in S, x \neq y$$

$$d(x, x) = 0 \quad \forall \quad x \in S$$

$$d(x, y) \leq d(x, z) + d(z, y) \quad \forall \quad x, y, z \in S$$

Definition
5.3.3

To *iterate* means to repeat, using the output of each iteration as the input for the next. The functional values obtained are called *iterates* of the function.

Definition
5.3.4

An iterated function system (IFS) consists of a finite set of functions, or mappings, and a metric space in which to operate.

The iterates of a given function may be written as $f(x)$, $f(f(x))$, $f(f(f(x)))$, and so on. For instance, the set of rectangles seen in Figure 5.3.1 was created by iterating the dilation D four times, using the largest rectangle as the initial input set.

$$D = \begin{bmatrix} .5 & 0 & 0 \\ 0 & .5 & 0 \\ 0 & 0 & 1 \end{bmatrix}$$

Figures 5.3.2–5.3.5 show successive iterations of an IFS consisting of four affine transformations. Each transformation takes a rectangle as input and returns rectangles as output. For instance, in Figure 5.3.2 the large rectangle that forms the border of the figure is mapped onto the four smaller rectangles appearing within the border. Each smaller rectangle is the image of the large rectangle under a different transformation. Collectively, the set of four transformations serves the same purpose as

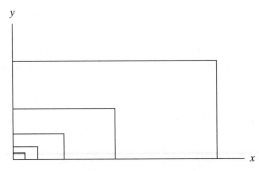

Figure 5.3.1 Iterated Dilation

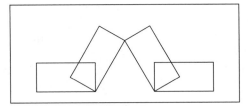

Figure 5.3.2 IFS 1

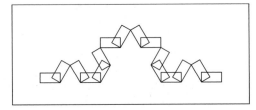

Figure 5.3.3 IFS 2

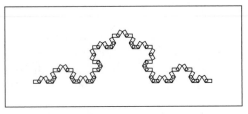

Figure 5.3.4 IFS 3

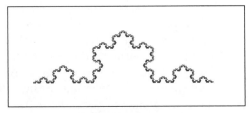

Figure 5.3.5 IFS 4

```
to ifs :n                          to line2 :n :l
pu ht rt 60 bk 120 rt 30 pd        ifelse :n=1 [box :l fd :l lt 60 box :l fd :l
make "x 1                          rt 120 box :l fd :l lt 60 box :l fd :l]
repeat :n [make "x 3*:x]           [line2 :n-1 :l lt 60 line2 :n-1 :l rt 120
make "l 250/:x                     line2 :n-1 :l lt 60 line2 :n-1 :l]
ifelse :n=0 [fd :l] [line2 :n :l]  end
pu home pd setfc
end                                to box :l
                                   repeat 4[fd :l lt 90 fd .5*:l lt 90 fd :l lt
                                   90 fd .5*:l lt 90]
                                   end
```

Table 5.3.1 IFS Concept Logo Code

the generators of section 5.2, replication of a given pattern on different scales. The smaller rectangles in Figure 5.3.2 are the output of the first iteration of the IFS.

In the second iteration, the same set of affine transformations is applied to each of the four smaller rectangles in Figure 5.3.2, producing the 12 rectangles in Figure 5.3.3. As this process is repeated again and again, successive rectangles become smaller and smaller. After a few iterations, the rectangles shrink to the size of a pixel, or screen element. From that point on, the computer appears to be plotting points and the image becomes indistinguishable from curves drawn using Logo (see Table 5.3.1). Systems of affine transformations such as this IFS may be used to create an enormous variety of fractals. The only requirement is that each transformation in the IFS must be a *contraction mapping*.

Definition 5.3.5

A function *f* is a *contraction mapping* if given a metric space (S, d) and a real constant $1 > c \geq 0$, then $c \cdot d(x, y) \geq d(f(x), f(y)) \; \forall \; x, y \in S$.

Figure 5.3.6 shows several examples of fractals created by iterated function systems. The affine transformations used to create each fractal are indicated by boxes representing the first level iterations of the IFS. The number of mappings in each IFS is therefore the same as the number of boxes shown.

In general, computer programs for creating fractals do not approach the task by systematically plotting smaller and smaller rectangles or any other geometric figure. They just plot points. For instance, Figure 5.3.7 shows an image from the shareware program *Chaos: The Software* by Josh Gordon, Rudy Rucker and John Walker. [Note the arrows added to locate the origin and coordinate axes.] The fractal, known as the Sierpinski gasket, is plotted one point at a time in a manner that could only be described as chaotic. Unlike the elegant, lock-step sequences of segments plotted using the *Geometers Sketchpad* and Logo, fractals created by iterated function systems emerge from a chaotic spray of points. One

URL 5.3.1

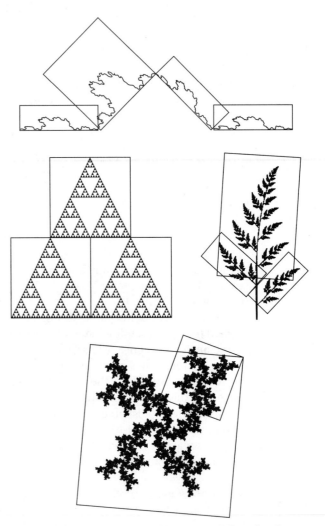

Figure 5.3.6 Fractals Created by Iterated Function Systems

of fractal geometry's most important messages is that chaotic processes do not always produce chaotic results. Indeed, there is something magical in the way the Sierpinski gasket emerges from the chaos of the software display. To see beyond the magic, some mathematics is required.

The three mappings used to create the Sierpinski gasket in Figure 5.3.7 are

$$\text{Map 1} = \begin{bmatrix} .5 & 0 & -1 \\ 0 & .5 & 1 \\ 0 & 0 & 1 \end{bmatrix} \quad \text{Map 2} = \begin{bmatrix} .5 & 0 & 1 \\ 0 & .5 & 1 \\ 0 & 0 & 1 \end{bmatrix} \quad \text{Map 3} = \begin{bmatrix} .5 & 0 & 0 \\ 0 & .5 & -1 \\ 0 & 0 & 1 \end{bmatrix}$$

Map 1 transforms the rectangle surrounding the Sierpinski gasket into the smaller rectangle superimposed on the upper left-hand portion

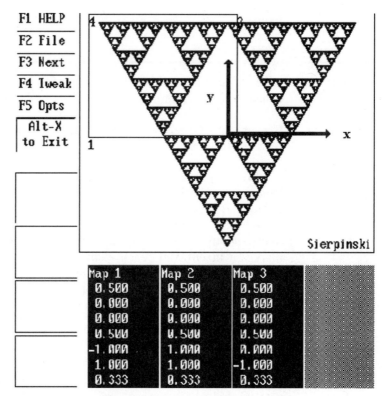

Figure 5.3.7 Sierpinski Gasket IFS

of the image. The other two maps produce similar rectangles in the upper right-hand portion and the bottom center portion of the image. Each of these dilations has a fixed point. Using techniques presented in Section 4.6, the fixed point of Map 1 may be determined to be $(-2, 2)$. The fixed points of Map 2 and Map 3 are $(2, 2)$ and $(0, -2)$, respectively.

The fractal seen in Figure 5.3.7 was created using the following algorithm:

1. As an initial point, randomly select and plot the fixed point of one of the mappings.
2. Randomly select one of the three mappings and apply it to the initial point to produce the second point.
3. Randomly select one of the three mappings and apply it to the second point to produce the third point.
4. Randomly select one of the three mappings and apply it to the third point to produce the fourth point.
5. Continue iterating until there are enough points in the image to characterize the fractal.

Using *Chaos: The Software*, this process occurs very fast. The speed at which the points are plotted obscures a significant fact: By randomly selecting the mapping at each iteration, each program run plots a different sequence of points. No two are identical. Nevertheless, the images created are clearly approximations of the same object, called the *strange attractor* of the iterated function system.

Definition 5.3.6

A *strange attractor* is a diagram or figure that characterizes the long term behavior of a chaotic system.

Theorem 5.3.1

If every function in a given IFS is a contraction mapping, then the strange attractor for the IFS is a fractal.

While Theorem 5.3.1 makes it simple to create fractals, the real challenge lies in creating fractals that are reminiscent of objects in the natural world or that possess interesting abstract features. The user interface for *Chaos: The Software* makes it easy to experiment with IFS mappings and to observe the effect on their associated strange attractors. In the window marked Current Map, each element may be individually increased or decreased. The effects of such changes are immediately plotted in the window to the left. In addition to the Sierpinski gasket, *Chaos: The Software* also includes a number of other interesting examples of iterated function systems, such as the fern fractal (see Figure 5.3.8).

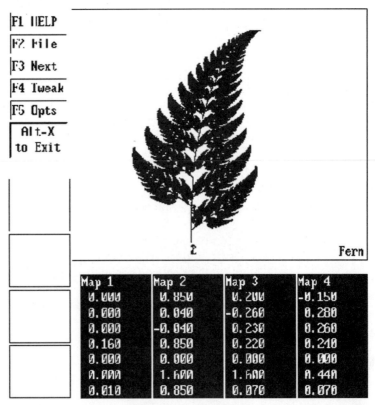

Map 1	Map 2	Map 3	Map 4
0.000	0.850	0.200	-0.150
0.000	0.040	-0.260	0.280
0.000	-0.040	0.230	0.260
0.160	0.850	0.220	0.240
0.000	0.000	0.000	0.000
0.000	1.600	1.600	0.440
0.010	0.850	0.070	0.070

Figure 5.3.8 Fern IFS

The IFS shown in Figure 5.3.8 has a strange attractor that resembles a fern. The four mappings of this IFS may be represented in matrix notation as follows:

$$\text{Map 1} = \begin{bmatrix} 0 & 0 & 0 \\ 0 & .16 & 0 \\ 0 & 0 & 1 \end{bmatrix} \qquad \text{Map 2} = \begin{bmatrix} .85 & .04 & 0 \\ -.04 & .85 & 1.6 \\ 0 & 0 & 1 \end{bmatrix}$$

$$\text{Map 3} = \begin{bmatrix} .20 & -.26 & 0 \\ .23 & .22 & 1.6 \\ 0 & 0 & 1 \end{bmatrix} \qquad \text{Map 4} = \begin{bmatrix} -.15 & .28 & 0 \\ .26 & .24 & .44 \\ 0 & 0 & 1 \end{bmatrix}$$

Modifying any element of these mappings is certain to introduce some change in the IFS. While the general nature of such effects may be inferred on the basis of considerations explored in Chapter Four and a study of dynamical systems, the precise details are generally not knowable without actually plotting the strange attractor. As a result, experimentation plays an important role in the study of such objects.

Figure 5.3.9 shows the strange attractor for an IFS consisting of 12 mappings, one for each rectangle superimposed on the strange attractor. This fractal was created using the java applet available at URL 5.3.2.

URL 5.3.2

Figure 5.3.9 Dave Fractal

Summary

If every mapping of an Iterated Function Systems is a contraction mapping, the strange attractor of the IFS is a fractal. While the reasons for this discovery are beyond the scope of this text, its power may be explored easily using computer software tools such as *Chaos: The Software*. Explorations conducted using this and other tools give us all an opportunity to personalize our knowledge of and experience with fractals.

Definitions	
5.3.1	A *space S* is a set. The elements of the set are the points of the space.
5.3.2	A *metric space* (S, d) consists of a space S and a function d that associates a real number with any two elements of S. The function d must have the following properties:

$$d(x, y) = d(y, x) \quad \forall \quad x, y \in S$$

$$0 < d(x, y) < \infty \quad \forall \quad x, y \in S, x \neq y$$

$$d(x, x) = 0 \quad \forall \quad x \in S$$

$$d(x, y) \leq d(x, z) + d(z, y) \quad \forall \quad x, y, z \in S$$

5.3.3	To *iterate* means to repeat, using the output of each iteration as the input for the next. The functional values obtained are called *iterates* of the function.
5.3.4	An iterated function system (IFS) consists of a finite set of functions, or mappings, and a metric space in which to operate.
5.3.5	A function f is a contraction mapping if given a metric space (S, d) and a real constant $1 > c \geq 0$, then $c \times d(x, y) \geq d(f(x), f(y)) \; \forall \; x, y \in S$.
5.3.6	A *strange attractor* is a diagram or figure that characterizes the long term behavior of an iterated function system.
Theorems	
5.3.1	If every function in a given IFS is a contraction mapping, then the strange attractor for the IFS is a fractal.

Table 5.3.2 Summary, Section 5.3

URLS		Note: Begin each URL with the prefix http://
5.3.1	s	www.mathcs.sjsu.edu/faculty/rucker/chaos.htm
5.3.2	s	http://www.dsv.su.se/~rogerh/ TheChaosGameIntroduction.html

Table 5.3.3 Section 5.3 URLs (c = concept, h = history, s = software, d = data)

Exercises

1. Show that the Euclidean plane is a metric space.
2. Show that each of the following linear transformation matrices is a contraction mapping.

a) $\begin{bmatrix} .5 & 0 & -1 \\ 0 & .5 & 1 \\ 0 & 0 & 1 \end{bmatrix}$
b) $\begin{bmatrix} .20 & -.26 & 0 \\ .23 & .22 & 1.6 \\ 0 & 0 & 1 \end{bmatrix}$

3. Write three linear transformation matrices that are not contraction mappings. Enter the matrices into *Chaos: The Software* and iterate the function system. Describe the result. Is it a strange attractor? Why?

4. Write three linear transformation matrices, each a composition of a contraction mapping and a reflection. Enter the matrices into *Chaos: The Software* and iterate the function system. Describe the result. Is it a strange attractor? Why?

5. Using *Chaos: The Software*, modify the Sierpinski Triangle IFS included as an example to resemble the following figure. Explain the changes that are necessary.

6. Starting with the Fern example in *Chaos: The Software*, modify the transformation matrices to produce a strange attractor that resembles a tree. Write the matrices and discuss the relationship between the matrices and the strange attractor.

7. Find the fixed point for each of the following mappings and a sequence of 10 iterates using $(1, 1, 1)$ as a seed point. How far is the 10th iterate from the mapping's fixed point?

$$\text{a)} \begin{bmatrix} .5 & 0 & -1 \\ 0 & .5 & 1 \\ 0 & 0 & 1 \end{bmatrix} \qquad \text{b)} \begin{bmatrix} .20 & -.26 & 0 \\ .23 & .22 & 1.6 \\ 0 & 0 & 1 \end{bmatrix}$$

8. Using Roger Holmberg's IFS applet located at the URL http://www.dsv.su.se/~rogerh/TheChaosGameIntroduction.html, create a fractal of your name similar to that shown in Figure 5.3.9.

5.4 From Order to Chaos

In this section, you will . . .

- **Investigate the periodic behavior of points under the mapping $f(x) = rx(1 - x)$, also known as the logistic map.**

Order and chaos have long been used as metaphors for other abstractions generally viewed as opposites of one another: good vs. evil, stability vs. instability, growth vs. decay, and so on. In this tradition, order and chaos are thought of as fundamentally incompatible, separate, different, and unrelated. One of the most surprising discoveries of twentieth century mathematics is that order and chaos are strongly related.

The Logistic Model

URL 5.4.1

Pierre Verhulst (1804–1849), a Belgian mathematician, studied mathematical models of population growth. Unlike his predecessors in the field, Verhulst did not believe that population growth followed a simple geometric progression in which the rate of growth was given by a fixed ratio r. Verhulst believed that environmental factors tend to prevent unlimited population growth. He accounted for such factors in a mathematical model $f(x) = rx(1 - x)$, also known as the *logistic map*. In this map, the variable x takes on values between 0 and 1, representing the current population as a decimal part of the environment's carrying capacity. For instance, an x value of 0.5 indicates that the current population is half of the environment's carrying capacity for that species. The variable r represents the rate of growth in the population from generation to generation in an environment offering no constraints or limits to growth. By including factors that account for the growth rate, r, and environmental limits, $(1 - x)$, Verhulst achieved a breakthrough in the study of population dynamics.

Verhulst's model of population dynamics is capable of demonstrating a variety of long-term outcomes: Extinction; convergence to a single, nonzero state; cyclic repetition of a finite number of nonzero states; and unpredictable oscillation between an infinite number of states. In the first two cases, the logistic model converges to a single, fixed state. For instance, a population might eventually stabilize at 60% of the environment's carrying capacity. In the third case, a population might alternate between two states, say 40% and 70% of the environment's carrying capacity. In the fourth case, a population might experience unpredictable variation.

The long-term behavior of the logistic map is often investigated using *web plots*. In Figure 5.4.1, two equations are shown: $f(x) = 2x(1 - x)$ and $y = x$. A point x_1 acts as a seed point for starting iterations of the map $f(x) = 2x(1 - x)$. A vertical line drawn through x_1 intersects the graph of $f(x) = 2x(1 - x)$ at $f(x_1)$. A horizontal line drawn through $f(x_1)$ intersects the graph of $y = x$ at a point x_2. This point has an x-coordinate equal to $f(x_1)$, the functional value returned by the first iteration. Repeating this process, the value of the second iteration is obtained by drawing a vertical line through x_2 and finding its intersection with the graph of $f(x) = 2x(1 - x)$. All subsequent iterations are determined using the same process. In the case of Figure 5.4.1, this process converges to the point at the intersection of the two graphs. Repeated experiments confirm that any seed point in the interval $(0, 1)$ produces the same result. The following examples examine the long-term behavior of the logistic map for several values of r.

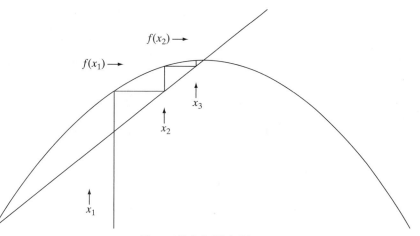

Figure 5.4.1 Web Plots

Example 5.4.1

Figure 5.4.2 shows a web plot and a table of values for the logistic map with $r = 1$. The table of values in the figure shows the first 20 iterates of the mapping, starting with the seed point $x_1 = 0.5$. The sequence of iterates beginning with the point x_1 is called the *orbit* of the point x_1. Both the web plot and the table of values suggest that the model converges to zero as the number of iterations goes to infinity. Zero is called a *fixed point* of the mapping. Because the orbit of every point in the interval $(0, 1)$ converges to this point, zero is called an *attracting fixed point* for the interval.

Definition 5.4.1

The sequence of iterates of a given point is called the *orbit* of the point.

Definition 5.4.2

A *fixed point* of a mapping is an invariant point of the mapping, that is, a point x such that $f(x) = x$.

Definition 5.4.3

An *attracting fixed point* is a point toward which other points tend under iteration. A *repelling fixed point* is a point away from which other points tend under iteration.

Example 5.4.2

Figure 5.4.3 shows a web plot and table of values for the logistic map with $r = 1.5$. Notice that the 19th and 20th iterates are identical to six decimal places. This strongly suggests that the decimal equivalent of $1/3$ is a fixed point of the mapping.

**URL
5.4.2**

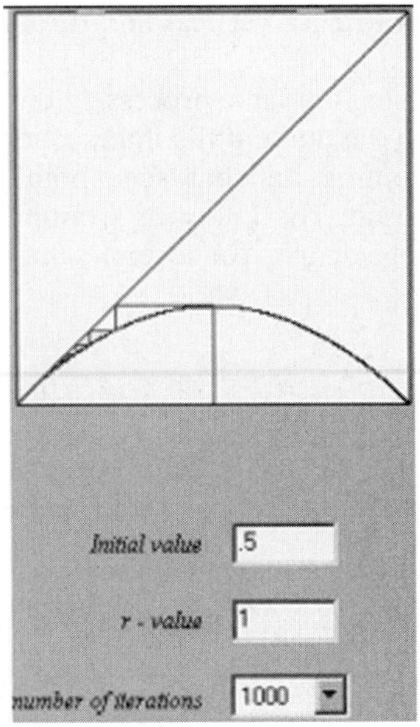

r	x	1-x	f(x)	Iteration
1	0.5	0.5	0.25	1
	0.25	0.75	0.1875	2
	0.1875	0.8125	0.152344	3
	0.152344	0.847656	0.129135	4
	0.129135	0.870865	0.112459	5
	0.112459	0.887541	0.099812	6
	0.099812	0.900188	0.08985	7
	0.08985	0.91015	0.081777	8
	0.081777	0.918223	0.075089	9
	0.075089	0.924911	0.069451	10
	0.069451	0.930549	0.064627	11
	0.064627	0.935373	0.060451	12
	0.060451	0.939549	0.056796	13
	0.056796	0.943204	0.053571	14
	0.053571	0.946429	0.050701	15
	0.050701	0.949299	0.04813	16
	0.04813	0.95187	0.045814	17
	0.045814	0.954186	0.043715	18
	0.043715	0.956285	0.041804	19
	0.041804	0.958196	0.040056	20

Figure 5.4.2 Logistic Map, $r = 1$

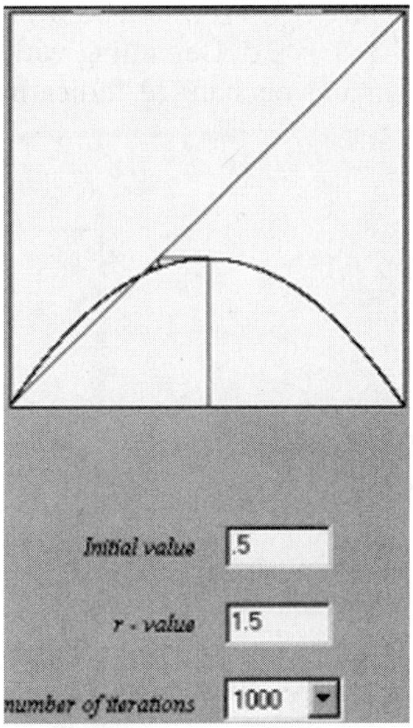

Initial value	.5
r - value	1.5
number of iterations	1000 ▾

r	x	1-x	f(x)	Iteration
1.5	0.5	0.5	0.375	1
	0.375	0.625	0.351563	2
	0.351563	0.648438	0.341949	3
	0.341949	0.658051	0.33753	4
	0.33753	0.66247	0.335405	5
	0.335405	0.664595	0.334363	6
	0.334363	0.665637	0.333847	7
	0.333847	0.666153	0.33359	8
	0.33359	0.66641	0.333461	9
	0.333461	0.666539	0.333397	10
	0.333397	0.666603	0.333365	11
	0.333365	0.666635	0.333349	12
	0.333349	0.666651	0.333341	13
	0.333341	0.666659	0.333337	14
	0.333337	0.666663	0.333335	15
	0.333335	0.666665	0.333334	16
	0.333334	0.666666	0.333334	17
	0.333334	0.666666	0.333334	18
	0.333334	0.666666	0.333333	19
	0.333333	0.666667	0.333333	20

Figure 5.4.3 Logistic Map, $r = 1.5$

Example 5.4.3   Figure 5.4.4 shows a web plot and table of values for the logistic map with $r = 2.5$. Beginning with the 17th iterate, the value of the mapping remains unchanged, indicating a fixed, or invariant, point.

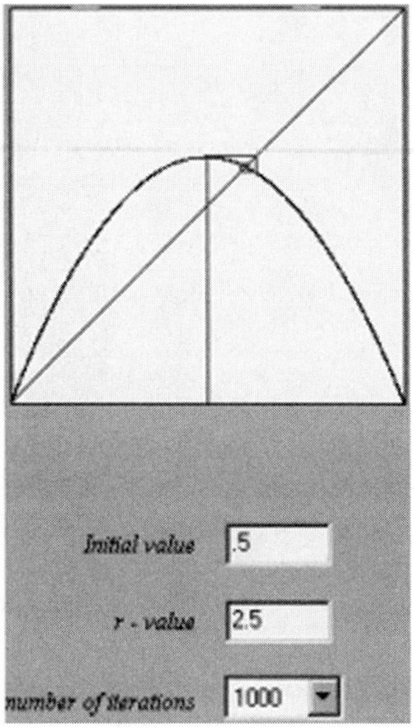

	Initial value	.5	
	r - value	2.5	
	number of iterations	1000 ▼	

r	x	1-x	f(x)	Iteration
2.5	0.5	0.5	0.625	1
	0.625	0.375	0.585938	2
	0.585938	0.414063	0.606537	3
	0.606537	0.393463	0.596625	4
	0.596625	0.403375	0.601659	5
	0.601659	0.398341	0.599164	6
	0.599164	0.400836	0.600416	7
	0.600416	0.399584	0.599791	8
	0.599791	0.400209	0.600104	9
	0.600104	0.399896	0.599948	10
	0.599948	0.400052	0.600026	11
	0.600026	0.399974	0.599987	12
	0.599987	0.400013	0.600007	13
	0.600007	0.399993	0.599997	14
	0.599997	0.400003	0.600002	15
	0.600002	0.399998	0.599999	16
	0.599999	0.400001	0.6	17
	0.6	0.4	0.6	18
	0.6	0.4	0.6	19
	0.6	0.4	0.6	20

Figure 5.4.4 Logistic Map for $r = 2.5$, Fixed Point

Example 5.4.4 Figure 5.4.5 shows a web plot and table of values for the logistic map with $r = 3.2$. Beginning with the 15th iterate, the model alternates between two states, 0.799455 and 0.513045. Considered as a set, these *periodic points* constitute a *periodic orbit* known as a *2-cycle*. Because all orbits eventually lead to this 2-cycle, it is called an *attracting periodic orbit*.

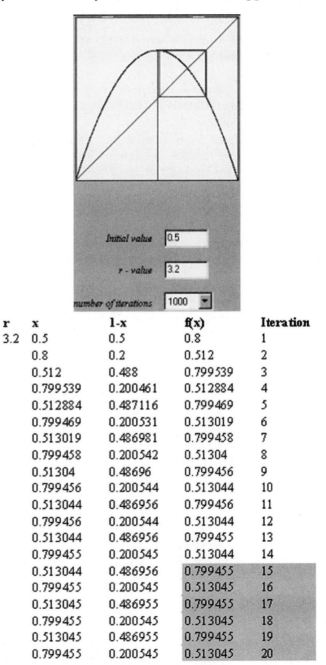

r	x	1-x	f(x)	Iteration
3.2	0.5	0.5	0.8	1
	0.8	0.2	0.512	2
	0.512	0.488	0.799539	3
	0.799539	0.200461	0.512884	4
	0.512884	0.487116	0.799469	5
	0.799469	0.200531	0.513019	6
	0.513019	0.486981	0.799458	7
	0.799458	0.200542	0.51304	8
	0.51304	0.48696	0.799456	9
	0.799456	0.200544	0.513044	10
	0.513044	0.486956	0.799456	11
	0.799456	0.200544	0.513044	12
	0.513044	0.486956	0.799455	13
	0.799455	0.200545	0.513044	14
	0.513044	0.486956	0.799455	15
	0.799455	0.200545	0.513045	16
	0.513045	0.486955	0.799455	17
	0.799455	0.200545	0.513045	18
	0.513045	0.486955	0.799455	19
	0.799455	0.200545	0.513045	20

Figure 5.4.5 Logistic Map, $r = 3.2$

Definition 5.4.4
A *periodic orbit* is finite set of points that, under iteration, returns to where it began. The *period* of the orbit is the same as the number of repeating points in the orbit.

Definition 5.4.5
The points of a periodic orbit are called *periodic points.*

Definition 5.4.6
An *n-cycle*, where *n* is a positive integer, is a set of *n* periodic points. For instance, a 4-cycle is a set of four periodic points.

Definition 5.4.7
An *attracting periodic orbit* is a cycle toward which other points tend under iteration of the mapping.

Example 5.4.5
Figure 5.4.6 shows a web plot and table of values for the logistic map with $r = 3.5$. Beginning with the fifth iterate, the model alternates between four states, 0.874997, 0.38282, 0826941, and 0.500884. These values constitute a 4-cycle.

Using a spreadsheet program such as *MS Excel*, the periodic behavior of points may be investigated by computing successive iterates of the mapping. For instance, Tables 5.4.1 and 5.4.2 show the approach used to create the numerical data in Figures 5.4.2–5.4.6.

r	*x*	$1 - x$	$f(x)$	**Iteration**
3.2	0.2	0.8	0.512	1
	0.512	0.488	0.799539	2
	0.799539	0.200461	0.512884	3

Table 5.4.1 Spreadsheet Values

r	*x*	$1 - x$	$f(x)$	**Iteration**
3.2	0.2	$= 1 - D5$	$= C5*D5*E5$	1
	$= F5$	$= 1 - D6$	$= \$C\$5*D6*E6$	$= G5 + 1$

Table 5.4.2 Spreadsheet Formulas

Examples 5.4.1–5.4.5 suggest that, in the long run, the logistic model may converge to a fixed value or alternate between a set of values. For instance, as *r* increases, the number of elements in these sets increases from 2 to 4 to 8... to 2^n. Changes in behavior such as these are called *period doubling bifurcations*. The specific values of *r* (rounded to seven decimal places) at which the bifurcations occur are listed after Figure 5.4.7. Figure 5.4.7, called a *bifurcation diagram*, shows the increasing complexity of the long-term behavior of the logistic map as *r* increases from 2.8 to 4.0.

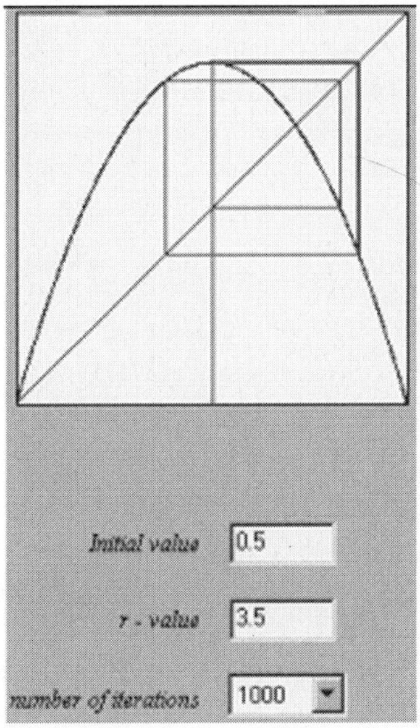

r	x	1-x	f(x)	Iteration
3.5	0.5	0.5	0.875	1
	0.875	0.125	0.382813	2
	0.382813	0.617188	0.826935	3
	0.826935	0.173065	0.500898	4
	0.500898	0.499102	0.874997	5
	0.874997	0.125003	0.38282	6
	0.38282	0.61718	0.826941	7
	0.826941	0.173059	0.500884	8
	0.500884	0.499116	0.874997	9
	0.874997	0.125003	0.38282	10
	0.38282	0.61718	0.826941	11
	0.826941	0.173059	0.500884	12
	0.500884	0.499116	0.874997	13
	0.874997	0.125003	0.38282	14
	0.38282	0.61718	0.826941	15
	0.826941	0.173059	0.500884	16
	0.500884	0.499116	0.874997	17
	0.874997	0.125003	0.38282	18
	0.38282	0.61718	0.826941	19
	0.826941	0.173059	0.500884	20

Figure 5.4.6 Logistic Map, $r = 3.5$

URL
5.4.3

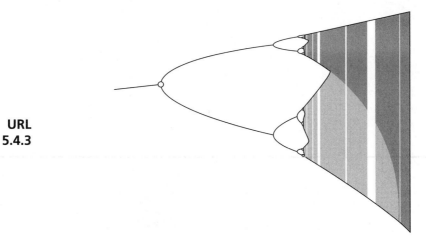

Figure 5.4.7 Bifurcation Diagram

**Definition
5.4.8** A *bifurcation* is a split or division that results in a fundamental change in behavior.

Bifurcation	r-value
1	3.000000
2	3.449489
3	3.544090
4	3.564407
5	3.568759

From Order to Chaos

As shown in Table 5.4.3, the bifurcation points get closer and closer together as r increases. At $r = 3.5699456$ something remarkable happens: The long-term behavior of the model becomes chaotic. For most values of r between 3.5699456 and 4.0, iterations range over a subset of the interval $(0, 1)$ in an unpredictable, non-periodic manner. This is evi-

r-value	Description	Behavior
$1 < r < 3$	Fixed Point Attractor	Convergence to a single point
$3 \le r < 3.5699456$	Periodic Orbit	Cyclic orbit with 2^n discrete states
$r \ge 3.5699456$	Chaos	Most values of r lead to unpredictable, non-periodic behavior
	Periodic Orbits	Some values of r lead to periodic orbits of length $k2^n$, where k is odd

Table 5.4.3 Long-term Behavior of the Logistic Model

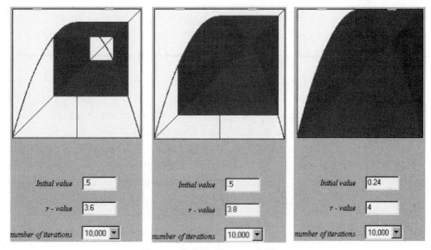

Figure 5.4.8 Onset of Chaos

dent in Figure 5.4.7 by the nearly solid column of points plotted for most values of r and in Figure 5.4.8 by the manner in which the web diagram appears to fill various subsets of the interval $(0, 1)$. For values of r greater than 4, iterates may take on any positive value.

Figure 5.4.7 contains narrow regions where few functional values appear. These are known as *periodic windows*. The widest such window is the period 3 window. Narrower windows to the left of the period 3 window are associated with cycles of period 5, 7, 9, 11, and so on. For instance, using an r value of 3.83, a 3-cycle with the following states is obtained: 0.957417, 0.156149, and 0.504666. A 5-cycle with states 0.934945, 0.227476, 0.657234, 0.842538, and 0.496176 is obtained using an r value of 3.74. Further investigation reveals additional periodic windows for cycles with length $k2^n$, where k is odd. Depending on the value of r, the long-term behavior of the model settles into one of the states in Table 5.4.3.

The structure of the bifurcation diagram shows a gradual transition from order to chaos in which periodic and chaotic orbits intermingle with one another. There is no clear demarcation, no "no man's land" between order and chaos. Indeed, the "distance" from order to chaos in terms of the parameter r may be very, very small. This phenomenon has consequences in the physical world. For instance, it is normal for the wind to induce oscillations in large mechanical structures such as bridges and television and radio antennas. An important design consideration is what happens to such oscillations over time. Are they naturally dampened by the structure or might they grow chaotically to unacceptable magnitudes, thereby damaging the structure? The study of chaos has led scientists and engineers to appreciate the interrelationship of order and chaos and to avoid designs that are inherently sensitive to small manufacturing variations from the design specifications or to perturbations that might drive the structure from stable to unstable performance.

Summary

Definitions	
5.4.1	The list of iterates of a given point is called the *orbit* of the point.
5.4.2	A *fixed point* is an invariant point of a given mapping $f(x)$, that is, $f(x) = x$.
5.4.3	An *attracting fixed point* is a point toward which points tend under iteration. A *repelling fixed point* is a point away from which points tend under iteration.
5.4.4	A *periodic orbit* is a finite set of points that repeats itself under iteration of a given mapping. The *period* of the orbit is the same as the number of repeating points in the cycle.
5.4.5	The points of a periodic orbit are called *periodic points*.
5.4.6	An *n-cycle*, where n is a positive integer, is a cycle whose period is n.
5.4.7	An *attracting periodic orbit* is a cycle toward which all points tend under iteration of the mapping.
5.4.8	A *bifurcation* is an abrupt qualitative change in behavior.

Table 5.4.4 Summary, Section 5.4

URLS		Note: Begin each URL with the prefix http://
5.4.1	h	www-history.mcs.st-and.ac.uk/history/Mathematicians/ Verhulst.html
5.4.2	s	www.math.fau.edu/kanser/Programs/ justbi.html
5.4.3	s	www.ukmail.org/~oswin/logistic.html

Table 5.4.5 Section 5.4 URLs (c = concept, h = history, s = software, d = data)

Exercises

1. Using procedures introduced in Chapter Four, find any fixed points of the following mappings. Determine whether any fixed points are attracting or repelling.
   a) $f(x) = x(1 - x)$
   b) $f(x) = 2x(1 - x)$
   c) $f(x) = 3x(1 - x)$
   d) $f(x) = 4x(1 - x)$

2. Use a spreadsheet to compute the orbits of the following points under the given mappings. If the points are periodic or eventually periodic, determine the periodicity of the orbit. If the orbit is chaotic, list the 100th iterate.

   a) $f(x) = x(1 - x); x_0 = .1$
   b) $f(x) = 2x(1 - x); x_0 = .2$
   c) $f(x) = 3x(1 - x); x_0 = .3$
   d) $f(x) = 4x(1 - x); x_0 = .4$

3. Using the applet at URL 5.4.2, plot a web diagram for each of the following mappings using the indicated seed point.

   a) $f(x) = x(1 - x); x_0 = .1$
   b) $f(x) = 2x(1 - x); x_0 = .2$
   c) $f(x) = 3x(1 - x); x_0 = .3$
   d) $f(x) = 4x(1 - x); x_0 = .4$

5.5 The Mandelbrot Set

- **Investigate the nature and features of Julia sets;**
- **Explore the relationship between the Mandelbrot set and Julia sets;**
- **Use spreadsheets to explore the periodic behavior of points located in different bulbs of the Mandelbrot set.**

In section 5.3, computer software tools are used to iterate sets of linear transformations enough times to imagine the graphical form of the strange attractor, a true fractal. The most famous of all fractals is the strange attractor known as the Mandelbrot set, which is created by iterating a quadratic mapping in the complex plane.

Julia Sets

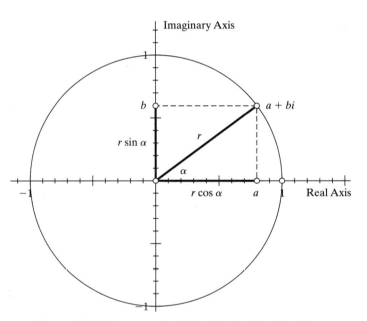

Figure 5.5.1 Representing Points in the Complex Plane

The complex plane, C, consists of all points of the form $a + bi$, where $a, b \in R$ and $i = \sqrt{-1}$. The distance, r, from the origin to the point $a + bi$ is known as the *modulus* of the point. In Figure 5.5.1, a point $a + bi$ is located on a unit circle, so its modulus is r. Using right triangle trigonometry, the position of this point may also be represented using the notation $r(\cos \alpha) + r(\sin \alpha)i$. Both notations are useful in exploring the Mandelbrot set.

Definition
5.5.1
The distance from the origin to a point in the complex plane is called the *modulus* of the point.

URL
5.5.1

Much of Mandelbrot's early work with fractals was focused on extending the work of a turn of the century mathematician named Gaston Julia. Julia was interested in iterated functions, including functions of complex variables. For instance, consider the mapping $z \rightarrow z^2$, where $z \in C$. This function maps the complex plane onto itself by squaring the coordinates of each point in the plane. Julia was interested in the long-term behavior of points in the complex plane under this transformation.

Using the two notations introduced above, the mapping $z \rightarrow z^2$ may be represented as

$$(a + bi)^2 = (a^2 - b^2) + (2ab)i, \quad \text{or as}$$

$$[r(\cos \alpha) + r(\sin \alpha)i]^2 = r^2[\cos 2\alpha + i \sin 2\alpha].$$

The second of these notations is more revealing. For instance, if $r > 1$, then $r^2 s > r$. Consequently, successive iterates move further and further from the origin. If $r < 1$, $r^2 < r$, so successive iterates move toward the origin. If $r = 1$, $r^2 = r$ and all iterates remain on the unit circle. In addition to motion toward or away from the origin, the angle formed by the modulus r and the x-axis is doubled on each iteration. The combined effect of these two motions may be visualized as a counterclockwise swirling action toward or away from the origin.

If you visualize successive iterations of this mapping, all of the points in the interior of the unit circle appear to swirl toward the origin, a mathematical "black hole" of sorts. The points outside the unit circle appear to swirl off to infinity. In other words, all of the points on the interior and exterior of the unit circle appear to be moving away from the circle. As a result, the circle is called a *repelling set* for the mapping. The union of all of the repelling points for a given mapping is called the *Julia set* of the mapping.

Definition
5.5.2

The union of all of the repelling points for a given mapping is called the *Julia set* of the mapping.

URL
5.5.2

For mapping of the form $z \rightarrow z^2 + c$, where c is a complex number held constant for any given mapping, the Julia sets obtained may be highly complex (see Table 5.5.1).

As seen in Table 5.5.1, Julia sets may take the form of complex curves or scattered points. In the case of complex curves, the Julia set is said to be *path-connected*, that is, there is a continuous path within the set between any two points of the set. In the case of scattered points, the Julia set is sometimes called *fractal dust*. While investigating which values of the parameter c lead to connected Julia sets, Benoit Mandelbrot discovered the set that made him famous and which now bears his name.

Definition
5.5.3

A set is said to be *path-connected* if any two points in the set can be joined by a curve lying entirely in the set.

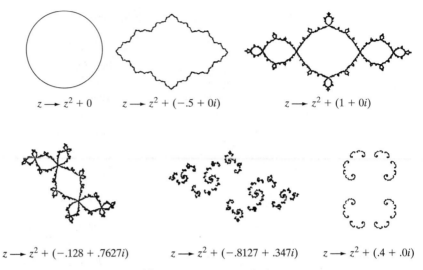

$z \longrightarrow z^2 + 0$ $z \longrightarrow z^2 + (-.5 + 0i)$ $z \longrightarrow z^2 + (1 + 0i)$

$z \longrightarrow z^2 + (-.128 + .7627i)$ $z \longrightarrow z^2 + (-.8127 + .347i)$ $z \longrightarrow z^2 + (.4 + .0i)$

Table 5.5.1 Sample Julia Sets

The Mandelbrot Set

Mandelbrot investigated the values of the parameter c for which the mapping $z \to z^2 + c$ has a path-connected Julia set, using $z = 0 + 0i$ as the seed point for the iteration process. In 1978, he developed a computer program that plotted the set of all points, c, in the *parameter plane* satisfying this criterion. Today that set is called the Mandelbrot set and is shown in Figure 5.5.2. In Figure 5.5.3 the coordinate axes of the complex plane are superimposed on the Mandelbrot set to indicate the position of the origin.

Figure 5.5.2 Mandelbrot Set

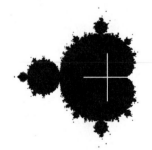

Figure 5.5.3 Mandelbrot Set and Coordinate Axes

Definition 5.5.4   The *Mandelbrot set* consists of all points c in the parameter plane for which the mapping $z \to z^2 + c$ in the complex plane has a path-connected Julia set.

URL 5.5.3   The first images of the Mandelbrot set showed a vaguely heart-shaped object covered with "buds" similar in shape to the main body of the object. With advances in computer graphics, views of the Mandelbrot set became available revealing details of its structure that motivated many mathematicians to focus their research efforts on the dynamics that give rise to the set. Figure 5.5.4 shows "zooms" on the Mandelbrot set, plotted in black, and nearby points, plotted in shades of gray.

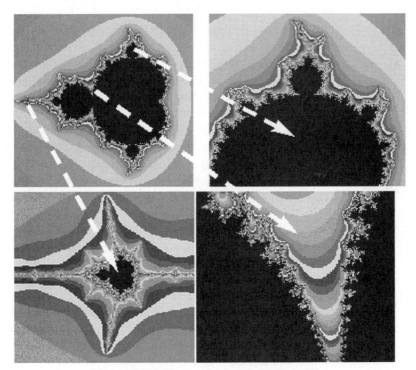

Figure 5.5.4 Zooms on the Mandelbrot Set

One of the most commonly used algorithms for plotting points in the Mandelbrot set is based on the discovery that the Julia set associated with any given choice of the parameter c in the mapping $z \rightarrow z^2 + c$ is never path-connected if the orbit of $z_0 = 0 + 0i$ (also called the *critical orbit*) tends to infinity and is always path-connected if the critical orbit is bounded. This *fundamental dichotomy* suggests a simple test for whether a given point c in the parameter plane is a member of the Mandebrot set. Iterate the mapping $z \rightarrow z^2 + c$ starting with $z_0 = 0 + 0i$. This process generates the sequence: $c, (c^2 + c), (c^2 + c)^2 + c, ((c^2 + c)^2 + c)^2 + c$. If, after many iterations of the mapping, the orbit tends to infinity, we assume that the Julia set for the mapping is not path-connected and the point c is not in the Mandelbrot set. For instance, if $c = 2 + 0i$, this sequence becomes 2, 6, 38, 1446. On the other hand, if the critical orbit remains bounded, we assume that the Julia set for the mapping is path-connected and that the point c is a member of the Mandelbrot set. For example, if $c = 0 + 1i$, the sequence $i, -1 + i, -i, -1 + i$ is obtained, a bounded orbit of period 2.

Which c-values do not belong to the Mandelbrot set? By using the triangle inequality, we can rule out a great many c-values on the basis of their distance from the origin. Let c be any complex number located 2 units or more from the origin, $|c| = 2 + d$ with $d > 0$, and let z be any point such that $|z| \geq |c|$. Then $|f(z)| = |z^2 + c| \geq |z^2| - |c| \geq |z|^2 - |z| = |z|(|z| - 1)$. Since $f(z) \geq |z|(|z| - 1) \geq |z|(1 + d)$ and since $(1 + d) > 1$, $|f(z)| \geq |z|$. In other words, outside a circle of radius two centered on the origin, $f(z)$ is always further from the origin than z. Consequently, the Mandelbrot set must lie within a circle of radius two centered on the origin. One need only check whether successive iterates of these points remain within two units of the origin to determine whether the sequence is bounded and therefore a part of the Mandelbrot set. In Figure 5.5.4, black points are in the Mandelbrot set. All other points (shades of grey) eventually escape.

Thanks to the efforts of many mathematicians, the structure of the Mandelbrot set and the dynamics that give rise to it are now well understood. One of the most interesting discoveries relates to the manner in which the Mandelbrot set catalogues the periodic behavior of points under successive iterations of the mapping $z \rightarrow z^2 + c$.

Periodicity and the Mandelbrot Set

The most important point in the Mandelbrot set, $0 + 0i$, is called the *critical point* of the mapping. This is the only point that is invariant under the mapping $z \rightarrow z^2 + c$. All other points in the Mandelbrot set vary from one iteration to the next. Under iteration, some points, c, immediately

CD
5.5.1

settle into periodic cycles of various lengths. For instance, if $c = -1 + 0i$, successive iterations alternate between the two points $-1 + 0i$ and $0 + 0i$, forming a cycle of period two. Some points eventually settle into a periodic cycle. For example, the seed point $-1.317 + 0i$ eventually settles into a 4-cycle consisting of the points $-1.1429527 + 0i$, $-0.0106591 + 0i$, $-1.3168864 + 0i$, $0.41718975 + 0i$. Tables 5.5.2 and 5.5.3 show several examples of this behavior. cd5_5_1 explores this behavior in the context of a *MS Excel* spreadsheet. Finally, some points follow chaotic orbits.

One of the most interesting aspects of the Mandelbrot set is the manner in which it catalogs the periodicity of points under the mapping $z \rightarrow z^2 + c$ by collecting points with the same periodicity into bulbs. As shown in Figure 5.5.5, all of the points in the main body of the set converge to cycles of length one. In the bulb immediately to the left of the main body, all of the points converge to cycles of length two. This is called the period 2 bulb. To the left of that is the period 4 bulb. A period 3 bulb is seen on the "top" of the set with a period 5 bulb to its left.

Length of Cycle	Seed Point	Periodic Orbit
1	$-.1 + 0i$	$0.11270167 + 0i$
2	$-1 + 0i$	$-1 + 0i$
		$0 + 0i$
3	$-.117 + .7583i$	$0.00600776 + -0.0275639i$
		$-0.1177237 + 0.75796881i$
		$-0.6776578 + 0.57983825i$
4	$-1.317 + 0i$	$-1.1429527 + 0i$
		$-0.0106591 + 0i$
		$-1.3168864 + 0i$
		$0.41718975 + 0i$
5	$-.5 + .575i$	$-0.8059914 + 0.38403453i$
		$0.00213966 + -0.0440571i$
		$-0.5019364 + 0.57481147i$
		$-0.578468 + -0.0020377i$
		$-0.1653789 + 0.57735743i$

Table 5.5.2 Periodic Orbits

C Real	C Imag	Iterate	Current R	Current I	New R	New I
-1	0	1	0	0	-1	0
		2	-1	0	0	0
		3	0	0	-1	0
		4	-1	0	0	0
		5	0	0	-1	0
		6	-1	0	0	0
		7	0	0	-1	0
		8	-1	0	0	0
		9	0	0	-1	0
		10	-1	0	0	0

Table 5.5.3 Spreadsheet Model: Periodic Orbits

Every bulb in the Mandelbrot set catalogs the periodicity of its points in this manner.

In addition to cataloging the periodicity of its points, each bulb also provides a sort of label for the periodicity of points in the bulb. As seen in Figures 5.5.6–5.5.8, each bulb has a variety of "antennae" that sprout from the boundary of the bulb. These antennae are keys to the periodicity of points in the bulbs. In each case, the number of branches (including the stem) of antenna emerging from the tip of the bulb is the same as the periodicity of the points in the bulb. For instance, in

Figure 5.5.5 Periods of the Bulbs

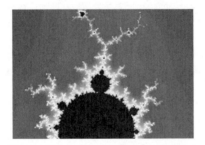

Figure 5.5.6 Period 3 Bulb Antenna

Figure 5.5.7 Period 4 Bulb Antenna

Figure 5.5.8 Period 5 Bulb Antenna

Figure 5.5.6, the main antenna consists of one stem and two branches, indicating a period 3 bulb. In Figure 5.5.7, the main antenna consists of one stem and three branches, indicating a period 4 bulb. Figure 5.5.8 shows a period 5 bulb.

Theorem 5.5.1 The period of each bulb in the Mandelbrot set is the same as the number of branches on the antennae sprouting from the tip of the bulb.

Summary

Mathematicians have discovered many aspects of the Mandelbrot set that may be interpreted as cataloging the behavior of points in many mappings. This amazing set is a kind of mathematical Rosetta Stone, explaining the dynamics of many mappings. Students interested in learning more about the Mandelbrot set and the dynamics that give rise to its structures are referred to the suggested readings at the end of this chapter.

Definitions	
5.5.1	The distance from the origin to a point in the complex plane is called the modulus of the point.
5.5.2	The union of all of the repelling points for a given mapping is called the Julia set of the mapping.
5.5.3	A set is said to be path-connected if any two points in the set can be joined by a curve lying entirely in the set.
5.5.4	The Mandelbrot set consists of all points c in the parameter plane for which the mapping $z \rightarrow z^2 + c$ in the complex plane has a connected Julia set.
Theorems	
5.5.1	The period of each bulb in the Mandelbrot set is the same as the number of branches on the antenna sprouting from the tip of the bulb.

Table 5.5.4 Summary, Section 5.5

URLs		Note: Begin each URL with the prefix http://
5.5.1	s	www-history.mcs.st-and.ac.uk/~history/Mathematicians/Julia.html
5.5.2	s	www.unca.edu/~mcmcclur/java/Julia/
5.5.3	s	www.softlab.ece.ntua.gr/miscellaneous/mandel/mandel.html

Table 5.5.5 Section 5.5 URLs (c = concept, h = history, s = software, d = data)

Exercises

1. Show that $[r(\cos \alpha) + r(\sin \alpha)i]^2 = r^2[\cos 2\alpha + i \sin 2\alpha]$.
2. Using a calculator or a spreadsheet program find the first 10 iterates of the mapping $z \rightarrow z^2$, using the following points as z_0. Is there a pattern to the orbit?

   a) $.5 - .1i$ d) $\sqrt{2} + \sqrt{2}i$
   b) $1 + 0i$ e) $2 + i$
   c) $0 + i$

3. Using a calculator or a spreadsheet program find the first 10 iterates of the mapping $z \rightarrow z^2 + c$, using the point $0 + 0i$ as z_0 and the following as the parameter c. Is there a pattern to the orbit?

   a) $5 - .1i$ d) $\sqrt{2} + \sqrt{2}i$

   b) $1 + 0i$ e) $2 + i$

   c) $0 + i$

4. Under the mapping $z \rightarrow z^2 + c$, find the coordinates of a point C with period

   a) 6   b) 7   c) 8   d) 9

Investigations

Exploring the Mandelbrot Set

Tool(s) Mandelbrot Viewer Applet   *Data File(s) none*

Focus

Aaron Michael Cohen's *Mandelbrot Viewer* Applet is a terrific tool for exploring the Mandelbrot set. It is available on-line at the URL http://home.netcom.com/~alcohen/MANDELBROT/Mandelbrot.html

Tasks

1. Zoom in on several bulbs, using the antennae to identify the periodicity of the bulbs. Read Robert Devaney's discussion of the periodicity of the bulbs *How to count and how to add* (http://math.bu.edu/ DYSYS/FRACGEOM2/FRACGEOM2.html).

2. Do you think the Mandelbrot set is self-similar in the same sense as the Koch curve? Why or why not?

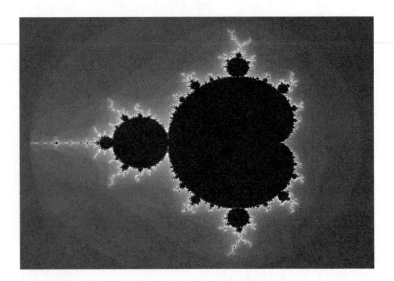

Exploring Julia Sets
Tool(s) Java Julia Set Generator   *Data File(s) none*

Focus
Mark McClure's *Java Julia Set Generator* is an excellent tool for exploring the Julia sets. It is available on-line at the URL http://www.unca.edu/~mcmcclur/java/Julia/

Tasks
1. Compare Julia sets for several pairs of points located on either side of the boundary of the Mandelbrot set. Describe the principal difference between the Julia sets for these points.

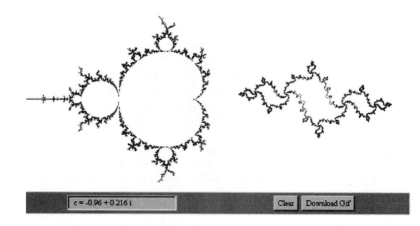

Suggested Readings

Barnsley, M. 1989. *Fractals everywhere.* Boston: Academic Press.

Bogomolny, A. *Plane filling curves.* Cut-the-knot, Inc. Available on-line http://cut-the-knot.com/do_you_know/hilbert.html

Bogomolny, A. *Fractal curves and dimension.* Cut-the-knot, Inc. Available on-line http://cut-the-knot.com/do_you_know/ dimension. html#sdim

Crilly, A., R. Earnshaw, and H. Jones, eds 1991. *Fractals and chaos.* New York: Springer-Verlag.

Devaney, R.L. ed. 1994. *Complex analytic dynamics: The mathematics behind the Mandelbrot and Julia Sets.* Providence: American Mathematical Society.

Devaney, R.L. 1989. *An Introduction to chaotic dynamical systems.* Addison-Wesley.

Devaney, R.L. 1989. *Chaos, fractals, and dynamics: Computer experiments in mathematics.* Menlo Park: Addison-Wesley.

Devaney, R.L. 1991. "The orbit diagram and the Mandelbrot Set." *The College Mathematics Journal,* 22, 23–38.

Dewdney, A.K. 1989, February. *Computer Recreations.* Scientific American, 108–111.

Fractal Coastlines. Center for Polymer Studies. Boston University. Available on-line http://polymer.bu.edu/java/java/coastline/ coastline.html

Geroges, J., D. Johnson, and R.L. Devaney, 1992. *A first course in chaotic dynamical systems.* Reading, MA: Addison-Wesley.

Gleick, J. 1987. *Chaos: Making a New Science.* New York: Viking Press.

Hausdorff dimension. Charles Sturt University, Australia. Available on-line http://life.csu.edu.au/fractop/doc/notes/chapt2a2.html

Holmberg, R. 1999. *The Chaos Game.* Available on-line http://www.dsv.su.se/~rogerh/TheChaosGameIntroduction.html

Jürgens, H., H.O. Peitgen, and D. Saupe, 1990, August. *The language of fractals.* Scientific American, 60-67.

Jürgens, H., H.O. Peitgen, and D. Saupe, 1992. *Chaos and fractals: New frontiers of science.* New York: Springer-Verlag.

Kanser, H.L. 1999. *Dynamical systems.* Available on-line http://www.math.fau.edu/kanser/Programs/justbi.html

Mandelbrot, B. 1967. *How long is the coast of Britain? Statistical self-similarity and fractional dimension.* Science, 155, 636–638.

Mandelbrot, B., 1982. *The Fractal geometry of nature.* San Francisco, Freeman & Co.

MSW Logo. Softronix, Inc. Available on-line http://www.softronix.com/

O'Connor, J.J. and E.F. Robertson, 1999. *The MacTutor History of Mathematics Archive.* School of Mathematics and Statistics, University of St. Andrews, Scotland. Available on-line:

Cantor. Available on-line http://www-history.mcs.st-and.ac.uk/ ~history/Mathematicians/Cantor.html

Hausdorff. Available on-line http://www-history.mcs.st-and.ac.uk/ ~history/Mathematicians/Hausdorff.html

Julia. Available on-line http:// www-history.mcs.st-and.ac.uk/ ~history/Mathematicians/Julia.html

Koch. Available on-line http:// www-history.mcs.st-and.ac.uk/ ~history/Mathematicians/Koch.html

Mandelbrot. Available on-line http://www-history.mcs.st-and.ac.uk/ ~history/Mathematicians/Mandelbrot.html

Peano. Available on-line http:// www-history.mcs.st-and.ac.uk/ ~history/Mathematicians/Peano.html

Sierpinski. Available on-line http://www-history.mcs.st-and.ac.uk/ ~history/Mathematicians/Sierpinski.html

Verhulst. Available on-line http://www-history.mcs.st-and.ac.uk/ ~history/Mathematicians/Verhulst.html

Oswin, S. Logistic function, or restricted and unrestricted growth function. Available on-line http://www.ukmail.org/~oswin/logistic.html

Peitgen H., Jürgens H. and D. Saupe 1992, "Length, area and dimension: Measuring complexity and scaling properties," in *Chaos and Fractals,* New York: Springer-Verlag, pp. 183–226.

Peitgen, H.-O., et al. 1991–1992. Fractals for the classroom Vols. 1 & 2. New York: Springer-Verlag.

Peitgen, H.-O., and Richter, P. 1986. *The beauty of fractals.* Berlin-New York: Springer-Verlag.

Peitgen, H.-O., and D. Saupe, 1988. *The science of fractal images.* Berlin-New York: Springer-Verlag.

Rucker, R. *Chaos: The Software download.* Available on-line http://www.mathcs. sjsu.edu/faculty/rucker/chaos.htm

Wright, D. 1996. *Fractal dimension.* Dynamical Systems and Fractals Lecture Notes Available on-line http://www.math.okstate.edu/mathdept/dynamics/lecnotes/ node37.html

Chapter 6

Projective Geometry

Of all the senses, sight is the most remarkable. At a glance, our optical-perceptual system scans millions of "bits" of information, selects some features for further analysis, stores or discards the rest, and constructs a meaningful scene from the objects it recognizes. This process is repeated dozens of times per second, producing a flood of information suitable for processing by the brain's other systems. Artists specialize in exploring and exploiting this remarkable system for our aesthetic and intellectual benefit.

Projective geometry was first developed by Renaissance artists and architects faced with a common problem: The geometric features of buildings and landscapes appear to change depending on one's point of view, or perspective. For instance, lines which are known to be parallel appear to converge in the distance (railroad tracks, highway margins, and so on). The near side of a rectangular wall appears to be longer than the far side. Steps at the bottom of a staircase appear wider than steps at the top. Distortions of this sort are an inevitable consequence of projecting a 3-dimensional world onto a 2-dimensional optical array, the human retina. Prior to the Renaissance, artists were familiar with the phenomena . . . but nobody had yet discovered the underlying mathematical basis for these effects or how to replicate them when representing the world in paintings and drawings. What the Renaissance artists needed and ultimately developed was a method for projecting 3-dimensional objects onto 2-dimensional surfaces. They called their method perspective drawing.

6.1 Elements of Perspective Drawing

In this section, you will . . .

- **Learn about map views and perspective views;**
- **Investigate the history of perspective drawing;**
- **Sketch simple objects in perspective.**

URL 6.1.1 Illustrations are used in professional and popular publications to present details and complex relationships that are difficult if not impossible to present using words alone. Illustrators, designers, architects, engineers, scientists, and other professionals rely on specialized computer software tools in the creation and presentation of both static and interactive graphic displays. For example, Figure 6.1.1 shows two perspective views

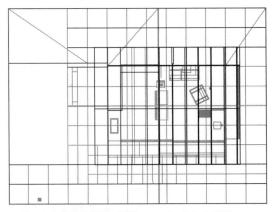

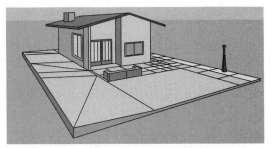

Figure 6.1.1 Three Views of a Model House

CD
6.1.1

and a top, plan view of a model house created using the architectural design tool *DesignWorkshop Lite*. The underlying mathematical basis for tools of this sort is called projective geometry. When applied to the arts, this basis is often called perspective drawing.

The same principles may be used to create realistic images of the natural world. For instance, Figure 6.1.2 shows a perspective view of a landscape, a map view of the data on which the scene is based, and a diagram illustrating the location of the camera and its field of view. The perspective view shows the lake and surrounding mountains as they would appear to an observer with the specified location and field of view. The apparent position, size, and elevation of every object in this scene are computed using mathematical transformations determined by the location and field of view of the observer. When the scene is rendered from a different location or with a different field of view, a different perspective view is obtained. Projective geometry provides a mathematical basis for understanding and specifying such transformations and for manipulating perspective views to achieve desired scientific and artistic results.

URL
6.1.2

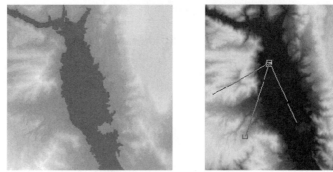

Perspective View of Landscape

Map View   Camera Position

Figure 6.1.2 Perspective View, Map View, and Camera Position

CD
6.1.2

The tool used to create the perspective view in Figure 6.1.2 is *Genesis II,* a freeware landscape rendering tool published by Geomantics, Ltd. The data on which the scene is based is a digital elevation model (DEM) file obtained from the United States Geological Survey (USGS).

The Origins of Perspective Drawing

URLs
6.1.3
&
6.1.4

Perspective drawing was developed in the Renaissance by an Italian architect, Leone Battista Alberti. Alberti was born in Genoa, Italy in 1404, second son of a wealthy merchant. Alberti acquired his early education at Gasparino Barzizza's Gymnasium in Padua, graduating in 1421. From there, he went to the University of Bologna, where he studied law but excelled in literature and geometry. After graduating in 1428, Alberti worked for several years as a secretary in the Papal Chancery in Rome, writing biographies of the saints in elegant Latin and traveling extensively in Europe. By 1432, he was living in Florence, working with famed artists Brunelleschi and Donatello. He also collaborated with Toscanelli on the development of maps later used by Columbus on his first voyage in 1492.

Of his many interests, Alberti was most enthusiastic about the development of a geometrical basis for perspective drawing. In 1434 he wrote the book, *Della Pittura*, the first written exposition on how to add a realistic third dimension to paintings. In it he said, "Nothing pleases me so much as mathematical investigations and demonstrations, especially when I can turn them to some useful practice drawing from mathematics the principles of painting perspective and some amazing propositions on the moving of weights." Alberti's influence on the development of Renaissance painting was significant and long-lasting. Leonardo da Vinci is known to have taken passages directly from *Della Pittura* and incorporated them into his *Trattato* (a common and legal practice at the time). He also greatly extended Alberti's ideas and techniques. Alberti's ideas did more than advance the development of painting, however. They revitalized the study of geometry. By introducing a new geometry, students of science were suddenly able to view Euclidean geometry as *a* geometry, rather than as *the* geometry. Few thinkers have enabled such a profound change in perspective. Alberti died in Rome in 1472.

Alberti's Method of Perspective Drawing

Alberti's method for creating a perspective view is illustrated in Figure 6.1.3. In this figure, the array of tiles seen in the Map View is converted into the array seen in the Perspective View. In the Perspective View, the eye is naturally drawn into the center of the image by the apparent convergence of parallel lines in the distance and by the realistic scaling of the tiles positioned different distances from the observer. Because there is only one vanishing point, the view is called 1-point perspective.

Alberti's objective was to convert top-down, or Map views, to Perspective views of a given scene. His procedure follows:

1. Position the objects in the composition in a rectangular window similar to that seen in the Map View, in this case a Euclidean view of the tile floor;
2. Draw line *m* parallel to and to the left of the window;
3. Represent the size and spacing of the tiles on line *m* by points A, B, C, D, and E;

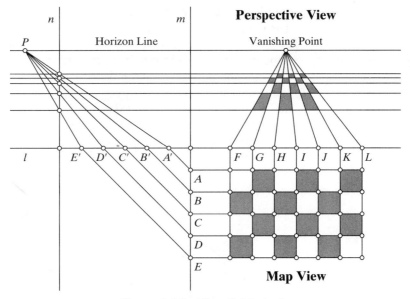

Figure 6.1.3 Alberti's Method

4. Construct line *l* perpendicular to line *m* above the Map View;
5. Represent the size and spacing of the tiles on line *l* by points *F, G, H, I, J, K,* and *L*;
6. Reproduce the spacing of points *A–E* on line *l* with points *A'–E'* and;
7. Construct line *n* parallel to line *m* and position it to the left of point *E'*;
8. Construct point *P* and segments *PA', PB', PC', PD',* and *PE'*;
9. Find the intersections of line *n* with segments *PA', PB', PC', PD',* and *PE'*;
10. Construct parallels to line *l* through each point of intersection;
11. Construct a parallel to line *l* through point *P*. This is the Horizon Line for the Perspective View;
12. Construct a point on the Horizon Line. This is the Vanishing Point for the Perspective View;
13. Position the Vanishing Point as shown and construct segments from the vanishing point to points *F, G, H, I, J, K,* and *L*;
14. Find the intersections of the lines defining the sides of each tile in the Perspective View;
15. Use the Polygon Interior option in the Construct menu in CD 6.1.3 to shade the tiles in the Perspective View.

CD
6.1.3

Once created using the *Geometers Sketchpad,* this model may be used to investigate the many perspectives made possible with Alberti's method. For instance, move point *P* along the Horizon Line. Move the Vanishing Point along the Horizon Line. Move line *n* left and right. How do these motions change the apparent perspective, or point-of-view, of the observer relative to the tile floor?

1-Point, 2-Point, and 3-Point Perspective

The goal of perspective drawing is to create a realistic illusion of depth in 2-dimensional illustrations. To do so, lines assumed to be parallel must appear to converge in the distance and objects assumed to be the same size must be drawn smaller and smaller according to their distance from the observer. These features are illustrated in the 2-dimensional black-and-white tiling pattern and the 5-sided prism shown in Figure 6.1.4. In these illustrations, the illusion of depth is achieved in one direction (front-to-back) without any corresponding sense of depth from side-to-side. This is an example of 1-Point Perspective.

CD
6.1.4
&
6.1.5

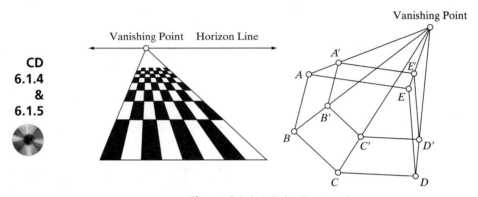

Figure 6.1.4 1-Point Perspective

Figure 6.1.5 shows an open cube using 2-point perspective. In 2-point perspective, parallel lines in two different planes converge to two different vanishing points. As a result, a sense of depth is created in two directions, making the object appear more 3-dimensional. Shading may be added to further enhance the illusion. For many artistic purposes, 2-point perspective is adequate.

CD
6.1.6

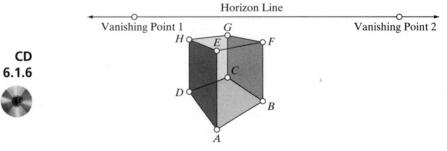

Figure 6.1.5 2-Point Perspective

A 3-point perspective view of a right rectangular prism is shown in Figure 6.1.6. If this were a view of an office building, the viewer would seem to be above and to the side of the building, looking down. Notice that there are three vanishing points.

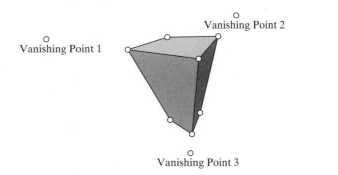

Figure 6.1.6 3-Point Perspective

Sketching a House in 2-Point Perspective

Basic sketching skills are useful to all mathematicians and a necessity for mathematics teachers. The following construction illustrates the use of 2-point perspective in drawing a simple building. These skills should be practiced until they are easy to use.

Step 1: Using the *Geometers Sketchpad,* draw a horizon line and select two vanishing points on the horizon line (see Figure 6.1.7). From each vanishing point, draw a line containing the base of a wall of the house. Mark the point of intersection $P1$. The two vanishing points and $P1$ determine the plane on which the house sits.

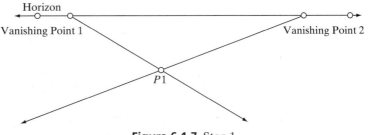

Figure 6.1.7 Step 1

Step 2: Draw a segment $P1$-$P2$ perpendicular to the horizon line. This determines the "front" edge of the house. Connect $P2$ to each vanishing point. This determines the plane of two of the walls of the house. Draw two other lines parallel to $P1$-$P2$ on either side of $P1$-$P2$, determining the ends of the wall planes, $P3$-$P4$ and $P5$-$P6$ (see Figure 6.1.8).

Step 3: Add hidden lines by extending additional rays from the vanishing points and label intersection points $P7$ and $P8$. Connect these points to the other corners as indicated and shade the faces to enhance the 3-dimensional appearance of the house (see Figure 6.1.9).

Step 4: Locate the centers ($P9$ and $P10$) of two opposite faces by finding the intersection of the diagonals of each face. Construct lines through $P9$ and $P10$ parallel to the vertical edges of the box. Select a point on the

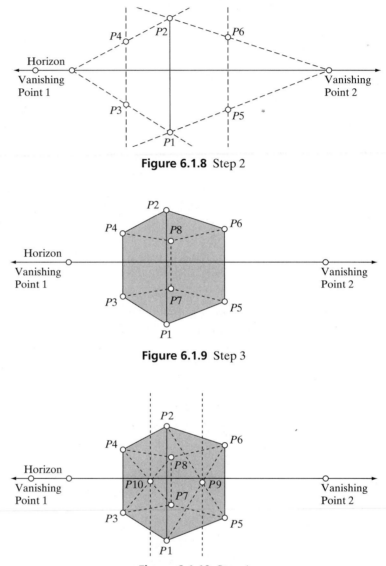

Figure 6.1.8 Step 2

Figure 6.1.9 Step 3

Figure 6.1.10 Step 4

nearer of the two lines to be the peak of the roof, labeling the point *P*11 (see Figure 6.1.10).

Step 5: Extend a line from *P*11 to Vanishing Point 1. Find the intersection of this line with the other line drawn parallel to the vertical edges of the box. Label the intersection *P*12. Connect *P*11 to *P*2 and *P*6 and *P*12 to *P*4 and *P*8 to make the roof (see Figure 6.1.11). Move *P*2 downward to achieve the perspective shown in Figure 6.1.11.

Step 6: Add additional shading, experiment with moving *P*2 and the two vanishing points, and hide lines and labels as desired (see Figure 6.1.12).

**CD
6.1.7**

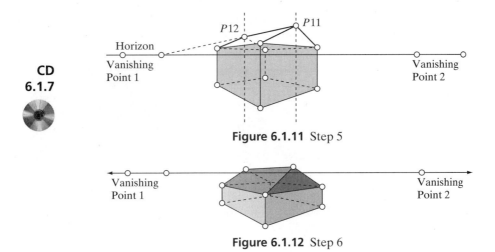

Figure 6.1.11 Step 5

Figure 6.1.12 Step 6

Summary

To the artists and architects of the Renaissance, Alberti's formulation of a geometric basis for perspective drawing was nothing less than revolutionary. Today, projective geometry provides a sound mathematical basis for a wide variety of computer-based visualizations in the arts, film and television, science, and engineering. While the geometric constructions introduced in this section are interesting in their own right, their analytic representations make perspective views possible in computer systems. The following sections explore those representations and their applications.

URLs		Note: Begin each URL with the prefix http://
6.1.1	s	www.artifice.com/dw_lite.html
6.1.2	s	www.geomantics.com/
6.1.3	h	www-history.mcs.st-andrews.ac.uk/history/ Mathematicians/Alberti.html
6.1.4	h	www.mega.it/eng/egui/pers/lbalber.htm

Table 6.1.1 Section 6.1 URLs (c = concept, h = history, s = software, d = data)

Exercises

1. Using Alberti's method, create a perspective view of the following floor tile pattern.

2. Using 1-point perspective, use the Geometers Sketchpad to do a sketch of each of the following objects.

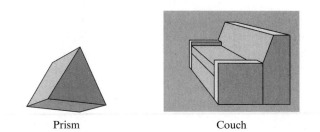

Prism   Couch

3. Using copies of illustrations from books and magazines, find two examples of each of the following. Locate all vanishing points implicit in each illustration.
   a) 1-point perspective
   b) 2-point perspective
   c) 3-point perspective
4. Take measurements on the following figure and enter your data into a table. Use the table to predict where the tops of the next two rows of tiles should be drawn. Draw two lines representing the tops of those rows. Do they appear to be "real?" Why or why not?

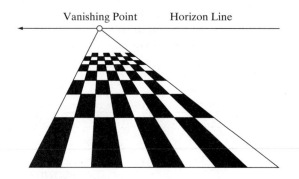

Investigations

Vanishing Points & Horizon Lines
Tool(s) Geometers Sketchpad   *Data File(s)* cd6_1_8.gsp

Focus
Parallel lines in the plane appear to converge at a distant point on the horizon.

Tasks
1. Move Vanishing Point #1 and Vanishing Point #2 along the horizon line. How do these motions appear to affect the position and size of the shaded quadrilateral relative to you?

2. Move the Horizon up and down the screen. How do these motions appear to affect the position and size of the shaded quadrilateral relative to you?
3. Change the separation between the lines by dragging points $a1$, $b1$, $a2$, and $b2$. How do these motions appear to affect the position and size of the shaded quadrilateral relative to you?
4. Position the vanishing points, horizon, and lines to create the impression that the shaded quadrilateral is close to your feet.

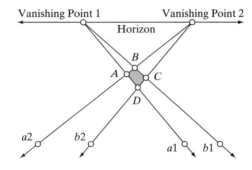

Distance Lines
Tool(s) Geometers Sketchpad   *Data File(s)* **cd6_1_4.gsp**

Focus
The apparent distance from the observer to the object is variable, as is the apparent height of the observer above the plane.

Tasks
1. Vary the position of the Distance Line. Describe the apparent effect on the shape and position of the object; on your height above the plane.
2. Vary the position of the Horizon. Describe the apparent effect on the shape and position of the object; on your height above the plane.
3. Vary the position of the Vanishing Point. Describe the apparent effect on the shape and position of the object; on your height above the plane.

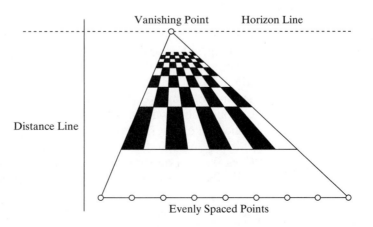

3-D Cube
Tool(s) Geometers Sketchpad   *Data File(s)* cd6_1_6.gsp

Focus
A dynamic, 2-point perspective representation of a cube.

Tasks
1. Move Vanishing Point 1 and Vanishing Point 2 along the Horizon
   Line. Move the drag point along the Horizon Line. Describe the
   apparent effect on the shape and position of the object.
2. Drag point *E* above the Horizon Line. Describe the apparent effect
   on the shape and position of the object.
3. Return point *E* to its original position and drag point *A* above the
   Horizon Line. Describe the apparent effect on the shape and position
   of the object.

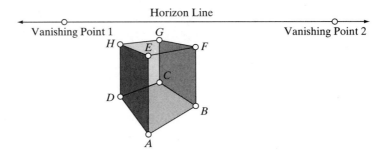

Two Views
Tool(s) Geometers Sketchpad   *Data File(s)* cd6_1_9.gsp,
                                        cd6_1_10.gsp, cd6_1_11.gsp

Focus
A 1-point perspective view of a right prism.

Tasks
1. Open all three files. Notice that cd6_1_9.gsp is an example. You will be
   working with cd6_1_10.gsp and cd6_1_11.gss.
2. Holding down the shift key, select six points in the order specified in
   cd6_1_11.gss.
   a) The point marked Origin in cd6_1_10.gsp
   b) Four points on the rectangular grid representing consecutive ver-
      tices of a quadrilateral.
   c) The point marked Height in cd6_1_10.gsp
3. Click the Fast button in cd6_1_11.gss.
4. Describe the features of the prism as represented in the 1-point perspective view.

5. Repeat the procedure without deleting the first prism. The second prism should be shorter than the first and positioned so as not to intersect the first prism. Color one prism red and the other blue.
6. Create a nested series of square prisms with bases of length 5, 4, and 3 units and heights of 1, 2, and 3 units respectively. Color and shade the prisms to better reveal their features and relationships to one another.
7. Create an object shaped like a set of stairs having 4 steps. Color and shade the result to better reveal the overall shape of the stairs.

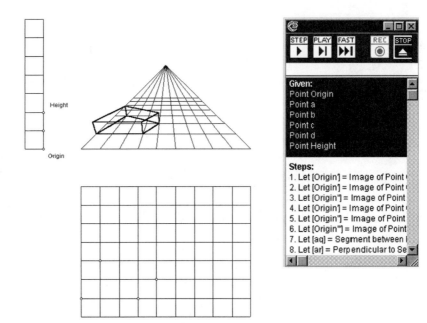

Alberti's Method
Tool(s) Geometers Sketchpad   *Data File(s)* cd6_1_3.gsp

Focus
Alberti's method for creating a 1-point perspective view.

Tasks
1. Vary the position of line *n*. Describe the apparent effects on the size and position of the object; on your position as observer.

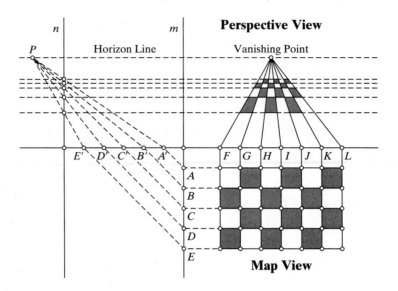

2. Vary the position of point *P*; of the vanishing point. Describe the apparent effects on the size and position of the object; on your position as observer.

3. This procedure creates an illusion of depth. Is that illusion true to reality or is it a false view of the world?

6.2 Introduction to Projective Geometry

In this section, you will . . .

- **Learn the axiomatic basis of projective geometry;**
- **Investigate examples of finite projective geometry;**
- **Apply finite projective geometry in error correcting codes.**

The development of perspective drawing in the Renaissance raised questions among scholars, artists, and architects concerning its relationship to Euclidean geometry. In time, a formal, axiomatic system was developed for this new branch of mathematics, now known as projective geometry.

Perspective drawing was developed with a single purpose in mind: Representing 3-dimensional objects on 2-dimensional surfaces. Ignoring matters of color, texture, and shade, perspective drawing creates a sense of realism by preserving some geometric features of objects and transforming other features. For instance, perspective drawing preserves collinearity of points and concurrence of lines. On the other hand, it does not preserve betweenness of points, or linear and angular measure (see Figure 6.2.1).

In Figure 6.2.1, a square is seen from two different perspectives (or points-of-view), Observer 1 and Observer 2. Since the square and both of the observers lie in the same plane, the image of square *ABCD* as seen from either perspective consists of a set of collinear points. Each set of image points is formed by the intersection of an image line with the lines connecting the observer's position and the vertices of the original

CD 6.2.1

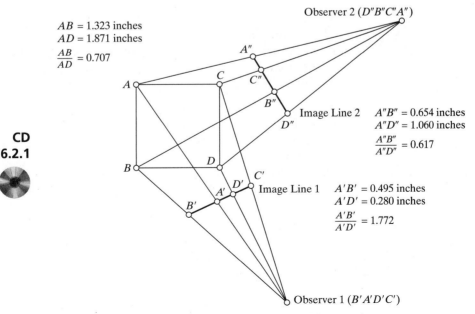

Observer 2 $(D''B''C''A'')$

$AB = 1.323$ inches
$AD = 1.871$ inches
$\frac{AB}{AD} = 0.707$

Image Line 2 $A''B'' = 0.654$ inches
$A''D'' = 1.060$ inches
$\frac{A''B''}{A''D''} = 0.617$

Image Line 1 $A'B' = 0.495$ inches
$A'D' = 0.280$ inches
$\frac{A'B'}{A'D'} = 1.772$

Observer 1 $(B'A'D'C')$

Figure 6.2.1 Betweenness and Perspective

square. Seen from the perspective of Observer 1, the vertices of the square are projected onto Image Line 1 with vertices ordered $B'A'D'C'$ from left to right. This projection positions point A' between points B' and D'. From the perspective of Observer 2, the vertices of the square are projected onto Image Line 2 with vertices ordered $D''B''C''A''$ from left to right. This projection positions point A'' between points C'' and D''. So, depending on the point-of-view of the observer, point A may appear to be between points B and D or points C and D. In general, the relative positions of points in a perspective view are dependent on the position of the observer. Considering that betweenness of points is not preserved, it is not surprising that linear and angular measurements are not preserved under perspective transformations (see Figure 6.2.1). Understanding precisely how betweenness and measurement depend on the observer's point of view is the central problem of perspective drawing. We begin with a comparison of points and lines in Euclidean and projective geometry.

Points and Lines
Points
- In the Euclidean plane, all points are ordinary in the sense that they represent locations that may be specified using ordered pairs of real numbers.
- In the projective plane, there are two sorts of points, ordinary points and "ideal" points (e.g., vanishing points). While artists and architects typically refer to ideal points as "points at infinity," mathematicians avoid that language because it implies that infinity is a point or location, neither of which is true.

Lines
- Lines in the Euclidean plane consist entirely of ordinary points.
- Lines in the projective plane contain both ordinary and ideal points. Given a line m, any line n parallel to m is concurrent with m at the ideal point, I_m. See Figure 6.2.2.

Parallel
- In the Euclidean plane, parallel lines have no points in common.
- In the projective plane, parallel lines have a single ideal point in common.
- Sets of parallel lines pointing in different directions have different ideal points. The set of all ideal points is called the ideal line.

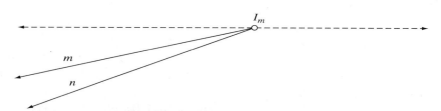

Figure 6.2.2 Ideal Points on a Line

These conventions are summarized in the following definitions:

**Definition
6.2.1**

Ordinary points are locations in the Euclidean plane (E) and may be represented using ordered pairs of real numbers.

**Definition
6.2.2**

Ideal points correspond to different directions in the Euclidean plane.

**Definition
6.2.3**

The set of all ideal points is called the *ideal line* (I) and corresponds to the set of all possible directions in the Euclidean plane.

**Definition
6.2.4**

The *real projective plane* consists of all points in the union $E \cup I$.

**Definition
6.2.5**

A set of points, all collinear, is called a *pencil* of points. The line containing the given points is called the *axis* of the pencil of points.

**Definition
6.2.6**

A set of lines, all concurrent, is called a *pencil of lines.* The point of concurrency is called the *center* of the pencil of lines.

**Example
6.2.1**

In Figure 6.2.3, collinear points $A, B, C,$ and D form a pencil of points with axis a. Concurrent lines $k, l, m,$ and n form a pencil of lines with center O.

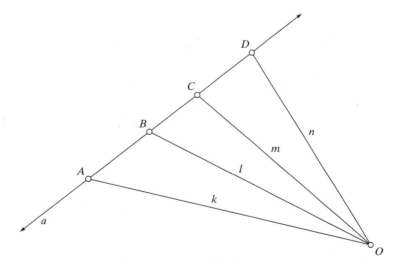

Figure 6.2.3 Pencils of Points and Lines

**Definition
6.2.7**

A *perspectivity* is a one-to-one mapping from a

- Pencil of lines L to a pencil of points P if each line in L is incident with exactly one point in P (see Figure 6.2.3.); or a
- Pencil of points P to a pencil of points P' if each line joining a point in P to its corresponding point in P' is incident with a fixed point O called the center of the perspectivity (see Figure 6.2.4); or a

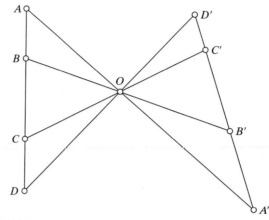

Figure 6.2.4 Pespectivity from a Pencil of Points to a Pencil of Points

- Pencil of lines O to a pencil of lines O' if the intersection of each line on O with its corresponding line in O' lies on a fixed line called the axis of the perspectivity (see Figure 6.2.5).

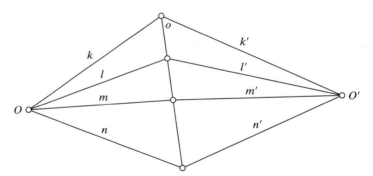

Figure 6.2.5 Pespectivity from a Pencil of Lines to a Pencil of Lines

Definition 6.2.8   A *complete quadrangle* consists of a set of four points, no three collinear, and the six lines determined by those points.

Definition 6.2.9   A *complete quadrilateral* consists of a set of four lines, no three concurrent, and the six points determined by those lines.

Example 6.2.2   In Figure 6.2.6, points A, B, E, D and lines $ABF, DEF, BEG, ADG, AEC, BDC$ constitute a complete quadrangle. Lines ABF, DEF, BEG, ADG and points A, B, D, E, F, G constitute a complete quadrilateral.

Theorem 6.2.1   Given any perspectivity P, an inverse perspectivity P' exists such that

$$PP' = P'P = I$$

Since all one-to-one mappings have inverses, and all perspectivities are one-to-one mappings, all perspectivities have inverses.

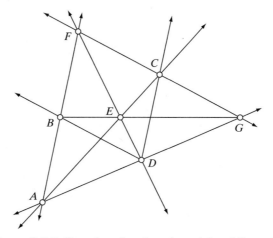

Figure 6.2.6 Complete Quadrangle and Quadrilateral

Axioms of Projective Geometry

As an extension of Euclidean geometry, projective geometry shares a number of features with Euclidean geometry. As shown in Table 6.2.1 there are also important differences.

PG1	There exists at least one line
PG2	Each line contains at least 3 points
PG3	Not all points are on the same line
PG4	Two distinct points determine a distinct line
PG5	Two distinct lines determine a distinct point

Table 6.2.1 Axioms of Projective Geometry

To facilitate investigation of this "new" geometry, mathematicians developed finite models in which to study these axioms and their logical consequences. While "tiny" geometries of this sort were developed to investigate the axiomatic nature of projective geometry, a number of interesting and powerful applications have emerged.

Finite Projective Geometry

In developing a finite projective geometry, a set of points is necessary, preferably small to simplify investigation. In addition, a new notion of "line" is needed, since Euclidean lines cannot exist in a finite geometry. These points and lines must satisfy each of the axioms for projective geometry given above. Figure 6.2.7 illustrates a model satisfying these requirements. For instance, line *ABF* satisfies axioms 1 and 2, consisting only of points *A*, *B*, and *F*. The Euclidean line containing these points is used to indicate that the three points constitute a line. A second line *DEF* is added satisfying axioms 3 and 5. By axioms 4 and 5, lines

BEG and *ADG* are necessary as well as their intersection, point *G*. By axiom 4, additional lines are necessary joining points *A* and *E*, *B* and *D*, and *F* and *G*. By axiom 2, each line must contain at least 3 points. So lines *AEC* and *BDC* are defined. To most students, line *BDC* seems different from the other lines, if not strange. It seems "bent." Since the "bend" occurs in the Euclidean line however, it is irrelevant. Recall that in this geometry, lines consist of 3-point sets with particular characteristics. *B*, *D*, and *C* constitute such a set and are therefore a line. Most importantly, the set of points and lines satisfies the axioms for a projective geometry.

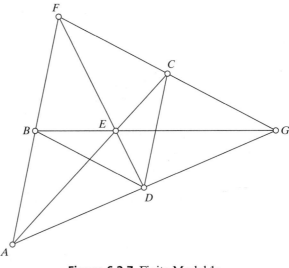

Figure 6.2.7 Finite Model 1a

Table 6.2.2 shows an alternative representation that avoids the use of Euclidean lines, focusing on the lines as sets. This model satisfies the axioms for projective geometry (see Table 6.2.3) and is called an order 2 finite projective plane.

Points	*A*	*B*	*C*	*D*	*E*	*F*	*G*
Lines	*A*	*B*	*F*	*D*	*B*	*A*	*A*
	D	*E*	*C*	*E*	*D*	*E*	*B*
	G	*G*	*G*	*F*	*C*	*C*	*F*

Table 6.2.2 Model 1b

A number of other observations of Model 1 are worth noting:

- The number of points on each line equals $2 + 1 = 3$.
- The number of lines on each point equals $2 + 1 = 3$.
- The number of points equals $2^2 + 2 + 1 = 7$.
- The number of lines equals $2^2 + 2 + 1 = 7$.

Axiom	Model 1 Characteristics in Table 6.2.2
PG1	For example, line l_1
PG2	Each column contains exactly three points
PG3	Point D is not on line l_1
PG4	Each pair of points occurs in exactly one column
PG5	Each pair of columns has exactly one point in common

Table 6.2.3 Model 1 Characteristics

Having verified that Model 1 satisfies the requirements of projective geometry, it is natural to ask whether other models having a greater number of points and lines are possible. Table 6.2.4 shows a model of an order 3 finite projective geometry. Table 6.2.5 demonstrates that Model 2 satisfies the axioms of a finite projective plane.

Points	A	B	C	D	E	F	G	H	I	J	K	L	M
Lines	A	A	A	A	B	B	B	C	C	C	D	D	D
	B	E	H	K	E	F	G	E	F	G	E	F	G
	C	F	I	L	H	I	J	I	J	H	J	H	I
	D	G	J	M	K	L	M	M	K	L	L	M	K

Table 6.2.4 Model 2

Axiom	Model 1 Characteristics in Table 6.2.4
PG1	For example, line $ABCD$
PG2	Each column contains exactly four points
PG3	Point E is not on line $ABCD$
PG4	Each pair of points occurs in exactly one column
PG5	Each pair of columns has exactly one point in common

Table 6.2.5 Model 2 Characteristics

Model 2 also satisfies the axioms of projective geometry. In addition,

- The number of points on each line equals $3 + 1 = 4$.
- The number of lines on each point equals $3 + 1 = 4$.
- The number of points equals $3^2 + 3 + 1 = 13$.
- The number of lines equals $3^2 + 3 + 1 = 13$.

When generalized, the features observed in Models 1 and 2 may be used to define a particular type of finite geometry.

Definition 6.2.10 Any set of points and lines satisfying the following axioms is called a *finite projective plane of order n.*

FPP1	There exist at least four distinct points, no three of which are collinear.
FPP2	There exists at least one line with exactly $n + 1$ $(n > 1)$ distinct points incident with it.
FPP3	Given two distinct points, there is exactly one line incident with both of them.
FPP4	Given two distinct lines there is at least one point incident with both of them.

Duality

The concept of duality involves both the structure and truth value of formal geometric statements. In projective geometry, the dual of any axiom or theorem is formed by replacing the word point with the word line and the word collinear with the word concurrent (and vice versa). In general, duals of Euclidean statements are not true. For instance, the dual of the Euclidean axiom, "Given any two distinct points, there is exactly one line incident with both of them" is the statement, "Given any two distinct lines, there is exactly one point incident with both of them." Since the dual statement disallows the possibility of parallel lines, it is not true. Table 6.2.6 shows the duals of the axioms for finite projective planes.

DPP1	There exist at least four distinct lines, no three of which are concurrent.
DPP2	There exists at least one point with exactly $n + 1$ $(n > 1)$ distinct lines incident with it.
DPP3	Given two distinct lines, there is exactly one point incident with both of them.
DPP4	Given two distinct points, there is at least one line incident with both of them.

Table 6.2.6 Duals of the Axioms for Finite Projective Planes

If the dual of each axiom in a given axiomatic system is true, then every theorem derived from the axioms would also be true in its dual form. Under these circumstances, the axiomatic system itself is said to have the property of duality. The following theorems formalize that concept in the case of projective geometry.

Theorem 6.2.2

There exists at least one point. [Dual of PG1]

By PG1 and PG2, there are at least three points. So, there is at least one point.

Theorem 6.2.3

Not all lines are on the same point. [Dual of PG3]

Assume that there is some point P that is on every line. Then for distinct lines m and n, $P \in m$ and $P \in n$. By PG2 each line contains at least two additional points. Let $A \in m$ and $B \in n$. By PG4, there is a line l containing A and B. Assume $P \in l$. Then points $A, B,$ and P are collinear. By PG4, there is only one line between A and P and B and P. Consequently, lines m and n are not distinct. This is a contradiction, so not all lines are on the same point.

Theorem 6.2.4

Each point is incident with at least three lines. [Dual of PG2]

By Theorem 6.2.3, not all lines are on a given point P. Let m be a line not incident with P. By PG2, line m contains at least three points. By PG4, each of those points and point P determine a distinct line. So, each point is incident with at least three lines.

Theorem 6.2.5

Two distinct lines determine a distinct point. [Dual of PG4]

This is the same as PG5

Theorem 6.2.6

Two distinct points determine a distinct line. [Dual of PG5]

This is the same as PG4.

Theorem 6.2.7

Given any line m, there is at least one point not on m.

By PG3, not all points are on the same line. So, there is at least one point not on m.

Theorem 6.2.8

In a finite projective plane, there is at least one point with exactly $n + 1$ lines incident with it $(n > 1)$. [Dual of FPP2]

By FPP2 there is a line m containing exactly $n + 1$ points. By FPP1 there is a point P not on m. By FPP3, there are $n + 1$ lines, l_i, joining point P and each of the $n + 1$ distinct points on line m. Because these $n + 1$ points are distinct, the l_i are also distinct. Assume that there is an additional line n through point P. By FPP4, line n must intersect line m at a point Q. Point Q cannot be any of the $n + 1$ points on line m or line n would not be distinct from the $n + 1$ lines, l_i. Then point Q must be an additional point on line m. This is a contradiction, so there are no additional lines through P.

Theorem 6.2.9

In a finite projective plane, each point is incident with exactly $n + 1$ lines $(n > 1)$.

Since the argument used in Theorem 6.2.8 uses an arbitrary point $P \notin m$, all points not on line m are incident with exactly $n + 1$ lines. So every point not on m is incident with exactly $n + 1$ lines. It is left as an exercise to show that all points on line m are also incident with exactly $n + 1$ lines.

Theorem 6.2.10	In a finite projective plane, each line is incident with exactly $n + 1$ points.

This statement is the dual of Theorem 6.2.9 and is true by duality.

Theorem 6.2.11	In a finite projective plane, there are four distinct lines, no three of which are concurrent. [Dual of FFP1]

By FPP1 and FPP3 there are at least two lines, m and n. By FPP4, lines m and n have a single point in common, P. By Theorem 6.2.10 there are at least two additional points per line, M_1, M_2, and N_1, N_2 respectively. No three of these points are collinear (why?). By FPP3, lines exist between the following pairs of points: M_1M_2, N_1N_2, M_1N_1, M_2N_2. Are any three of these four lines concurrent? Assume that lines M_1M_2, N_1N_2, and M_1N_1 are concurrent at a point Q. Since M_1M_2 and N_1N_2 are already concurrent at point P, point Q must be the same as point P (why?). If line M_1N_1 is incident with point P, then lines m and n cannot be distinct (why?). This is a contradiction. Then lines M_1M_2, N_1N_2, and M_1N_1 cannot be concurrent. As an exercise, rule out the possibility of any three of the other lines being concurrent.

Having seen finite projective planes of order 2 and 3, it is natural to wonder whether finite projective planes of other orders exist. Veblen and Bussey (1906) proved that finite projective planes of order n exist whenever n is a power of a prime. For example, finite projective planes exist of the following order: 2, 4, 8, 16 …; 3, 9, 27, 81 …; 5, 25, 125, 625 …; 7, 49, 343, 2401; and so on. Bruck and Ryser (1949) proved that if n is congruent to 1 or 2 (mod4) and if n cannot be written as the sum of two squares, then there are no projective planes of order n. For example, there are no finite projective planes of orders 6, 14, 21, and 22. This still leaves unresolved an infinite number of cases, however. As a consequence, the question, "For which values of n is there a finite projective plane of order n?" remains an unanswered question of mathematics.

URL 6.2.1	**Error Correcting Codes** For millions of people, digital telecommunications is an important element of their professional and personal lives. Digital signals encode information in a binary format, using groups of 0s and 1s to represent alphabetic, numeric, punctuation, and other symbols. In the early days of electronic computing, these groups became known as "words" or "bytes," each word consisting of eight binary "bits." For example, each of the following is an 8-bit binary word: 00000000, 00000001, 00000010, and 00000011. Since each bit has exactly two possible states, 0 or 1, there are $2^8 = 256$ possible 8-bit binary words. The smallest 8-bit binary number is $00000000_{two} = 0_{ten}$. The

largest 8-bit binary number is $11111111_{two} = 255_{ten}$. The columns in each word may be thought of as representing powers of two. For instance, the 8-bit binary number $01001101_{two} = (2^6 + 2^3 + 2^2 + 2^0)_{ten} = 75_{ten}$ (see Table 6.2.7). Combinations of these computer "words" are used to represent letters, numerals, punctuation, and other linguistic and mathematical symbols.

2^7	2^6	2^5	2^4	2^3	2^2	2^1	2^0
0	1	0	0	1	1	0	1

Table 6.2.7 Structure of Binary Code Words

Every day millions of people use electronic mail, WWW browsers, ATM machines, and satellite television services without giving a thought to how information is moved from one location to another. Most digital telecommunications systems transmit files over phone lines, fiber optics cables, or by radio and microwave signals. Modern technology has made these and other applications of digital telecommunications so reliable that it is rare even to hear of a transmission that involved an error. This reliability is based on a number of technologies, one of which is called error correcting codes.

Error correcting codes were first developed in the 1940s by Richard Hamming. Hamming was frustrated by shutdowns that occurred whenever a coding error was detected by his computer. As a result, he sought and found a strategy for detecting and correcting coding errors before they shut down the operation of his computer. One of the simplest examples of this strategy is known today as the Hamming (7,4) code. This code has as its mathematical basis an order 2 finite projective plane. Order 2 finite projective planes are known to have exactly seven points and seven lines. An alternative representation to that given in Table 6.2.2 is shown in Table 6.2.8 and is known as an incidence table.

A	1	0	0	0	0	1	1
B	0	1	0	0	1	0	1
C	0	0	1	0	1	1	0
D	1	0	0	1	1	0	0
E	0	1	0	1	0	1	0
F	0	0	1	1	0	0	1
G	1	1	1	0	0	0	0

Table 6.2.8 Incidence Table for an Order 2 Finite Projective Plane

The rows and columns of the incidence table suggest a way to specify either points or lines using ordered 7-tuples. For instance, point A may be represented as the 7-bit binary number 1001100. Because each of these 7-tuples contains exactly three 1s, we say that each has weight 3. This sort of representation led Hamming to explore the nature of single errors in 7-bit binary words.

Hamming designated sixteen of the $2^7 = 128$ available 7-tuples as code words. These code words are generated using the matrix G, where

$$G = \begin{bmatrix} 1 & 0 & 0 & 0 & 0 & 1 & 1 \\ 0 & 1 & 0 & 0 & 1 & 0 & 1 \\ 0 & 0 & 1 & 0 & 1 & 1 & 0 \\ 0 & 0 & 0 & 1 & 1 & 1 & 1 \end{bmatrix}$$

Notice that the first three rows of G correspond to the first three rows of Table 6.2.8. The fourth row is equivalent to the sum of the next four rows (mod2). Each code word listed in Table 6.2.9 represents the sum (mod2) of various combinations of the rows in matrix G. For example, the code word 1100110 is equivalent to the sum (mod2) of the first two rows of G. Similarly, the code word 1110000 is obtained by adding the first three rows of G. The code word 0000000 is obtained by adding any row to itself.

0000 000	1000 011
0001 111	1001 100
0010 110	1010 101
0011 001	1011 010
0100 101	1100 110
0101 010	1101 001
0110 011	1110 000
0111 100	1111 111

Table 6.2.9 Code Words

Like raisins in a muffin, these code words are spread throughout the 112 remaining 7-tuples in this seven dimensional space. Each code word is "surrounded" by non-code words, some of which differ from the code word in only one of the seven bit locations. Hamming's approach detects only those 7-tuples that differ from legitimate code words in only one bit location. Errors involving more than one digit location are not addressable using the Hamming (7,4) code.

Notice that each code word is broken down into a 4-tuple and a 3-tuple. The 4-tuples occupy the "information" positions in each word. This is the "content" portion of the code word. The 3-tuples occupy the "redundancy" positions. This is the portion of the code word that is used to detect and correct errors. In the Hamming (7,4) code, each incoming word is multiplied by a parity check matrix, P, where

$$P = \begin{bmatrix} 0 & 0 & 0 & 1 & 1 & 1 & 1 \\ 0 & 1 & 1 & 0 & 0 & 1 & 1 \\ 1 & 0 & 1 & 0 & 1 & 0 & 1 \end{bmatrix}$$

Multiplication of elements is computed (mod2). For instance,

$$\begin{bmatrix} 0 & 0 & 0 & 1 & 1 & 1 & 1 \\ 0 & 1 & 1 & 0 & 0 & 1 & 1 \\ 1 & 0 & 1 & 0 & 1 & 0 & 1 \end{bmatrix} \begin{bmatrix} 1 \\ 1 \\ 0 \\ 0 \\ 0 \\ 1 \\ 1 \end{bmatrix} = \begin{bmatrix} 0 \\ 1 \\ 0 \end{bmatrix}$$

The result $010_{two} = 2_{ten}$ indicates that there is an error in the second position from the top, or left, depending on how the binary number is written. Since each bit has only two values, 0 or 1, changing the bit value in the second position corrects the error. In this case the corrected word is 1000011. When the parity matrix is multiplied times a code word, the result is 000, indicating that there are no errors.

Because this approach creates only sixteen code words, a combination of code words is needed to uniquely identify all the symbols needed in computer and telecommunications applications. For many years, pairs of words were used to name symbols. This practice made $16 \times 16 = 256$ symbols available at one time: upper case letters, lower case letters, numerals 0–9, punctuation marks, mathematics symbols, and so on. Under this arrangement, each symbol was assigned a unique combination of two words. For instance, the lower case letter a is associated with the combination 01_{ten}.

When a computer or other communications device receives a message, each word in an incoming combination is checked using a strategy similar to the Hamming (7,4) code to make sure it is legitimate. If an error is detected, it may be corrected using a scheme similar to the Hamming (7,4) code or the computer that sent the non-code word may be asked to resend the message. In more elaborate error detecting and correcting schemes, additional bit locations are used in the redundancy positions. While this increases the "overhead" associated with sending and receiving information, it reduces the need to resend messages. In some contexts, such as satellite telemetry from deep space probes, this is a practical necessity, as the time involved in contacting the spacecraft to order a resend may be measured in hours.

Summary

Axiomatic projective geometry shares a number of features with Euclidean geometry. It also differs from Euclidean geometry in important ways, the most obvious being that, in projective geometry, all lines intersect. While perspective drawing is the most familiar application of real projective geometry, finite projective geometries provide a more convenient context in which to study projective geometry's logical structure and distinctive features, such as duality. By duality, each theorem has two forms, each obtained from the other by interchanging point with line and collinear with concurrent. For instance, in a finite projective plane, each line is on $n + 1$ points and each point is on $n + 1$ lines. Since Euclidean geometry is not dual, projective geometry is the first geometry studied by most students with this feature.

Finite geometries provide the mathematical basis for a number of applications in science and engineering. For instance, error correcting codes are used entensively in computer and digital telecommunications systems to detect and correct errors that, otherwise, would make electronic mail, banking, and other common applications unreliable.

URLs	Note: Begin each URL with the prefix http://
6.2.1	ch pass.maths.org.uk/issue3/codes/index.html

Table 6.2.10 Section 6.2 URL (c = concept, h = history, s = software, d = data)

Exercises

1. Sketch a perspectivity between a pencil of points and a pencil of lines.
2. Sketch a perspectivity between two pencils of points.
3. Sketch a perspectivity between two pencils of lines.
4. Determine whether the following systems are finite projective planes. Justify your answer.

a)

Points	*A*	*B*	*C*	*D*	*E*	*F*	*G*
Lines	*C*	*A*	*C*	*G*	*B*	*A*	*A*
	D	*E*	*F*	*E*	*F*	*G*	*F*
	B	*B*	*E*	*D*	*G*	*C*	*D*

b)

Points	1	2	3	4	5	6	7	8	9	10	11	12	13
Lines	1	2	3	4	5	6	7	8	9	10	11	12	13
	2	3	4	5	6	7	8	9	10	11	12	13	1
	4	5	6	7	8	9	10	11	12	13	1	2	3
	10	11	12	13	1	2	3	4	5	6	7	8	9

5. Determine whether the following are words in the Hamming (7,4) code and correct any errors.

a) $\begin{bmatrix} 0 \\ 0 \\ 1 \\ 0 \\ 0 \\ 0 \\ 0 \end{bmatrix}$ b) $\begin{bmatrix} 0 \\ 0 \\ 1 \\ 0 \\ 1 \\ 0 \\ 0 \end{bmatrix}$ c) $\begin{bmatrix} 0 \\ 1 \\ 1 \\ 0 \\ 0 \\ 1 \\ 1 \end{bmatrix}$

6. Which of the axioms of a projective plane are also true in Euclidean geometry? Which are not?
7. In an order-2 finite projective plane, how many different complete quadrangles are there? How many different complete quadrilaterals?
8. Does a finite projective plane of order 121 exist? Justify your answer.

Investigation

Betweenness of Points

Tool(s) Geometers Sketchpad   *Data File(s) cd6_2_1.gsp*

Focus
The order of vertices of a square as seen from different positions and the apparent division of space by the vertices.

Tasks
1. Within the range of motion allowed, does moving the position of the observer change the apparent order of the vertices?
2. What happens to the ratios $A'B'/A'D'$ and $A''B''/A''D''$ as the position of the observers changes?

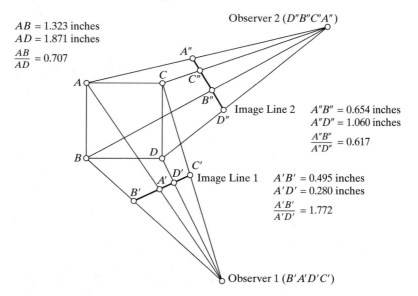

$AB = 1.323$ inches
$AD = 1.871$ inches
$\dfrac{AB}{AD} = 0.707$

Observer 2 $(D''B''C''A'')$

Image Line 2 $A''B'' = 0.654$ inches
           $A''D'' = 1.060$ inches
           $\dfrac{A''B''}{A''D''} = 0.617$

Image Line 1 $A'B' = 0.495$ inches
           $A'D' = 0.280$ inches
           $\dfrac{A'B'}{A'D'} = 1.772$

Observer 1 $(B'A'D'C')$

6.3 The Cross Ratio

In this section, you will . . .
- **Begin development of a vocabulary for perspective transformations;**
- **Investigate the concept of invariance in perspective transformations;**
- **Characterize the partition of a perspective view using the cross ratio.**

CD 6.3.1

Figure 6.3.1 shows a perspectivity between a pencil of lines (ℓ_1, ℓ_2, ℓ_3, and ℓ_4) with center P and a pencil of points ($ACBD$) with axis m. For an observer located at point P, points B and C partition the field of view defined by $\angle APD$. A mathematical treatment of perspective requires a well-defined method for characterizing that partition. Three approaches are presented based on the

1. Angles formed by line segments ℓ_1, ℓ_2, ℓ_3, and ℓ_4
2. Segments they cut off on line AD, d_1, d_2, and d_3, and
3. Slopes of segments ℓ_1, ℓ_2, ℓ_3 and ℓ_4.

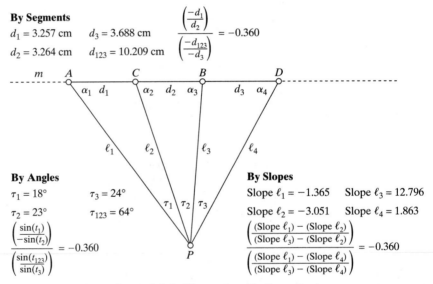

By Segments

$d_1 = 3.257$ cm $d_3 = 3.688$ cm

$d_2 = 3.264$ cm $d_{123} = 10.209$ cm

$$\frac{\left(\frac{-d_1}{d_2}\right)}{\left(\frac{-d_{123}}{-d_3}\right)} = -0.360$$

By Angles

$\tau_1 = 18°$ $\tau_3 = 24°$

$\tau_2 = 23°$ $\tau_{123} = 64°$

$$\frac{\left(\frac{\sin(t_1)}{-\sin(t_2)}\right)}{\left(\frac{\sin(t_{123})}{\sin(t_3)}\right)} = -0.360$$

By Slopes

Slope $\ell_1 = -1.365$ Slope $\ell_3 = 12.796$

Slope $\ell_2 = -3.051$ Slope $\ell_4 = 1.863$

$$\frac{\left(\frac{(\text{Slope }\ell_1) - (\text{Slope }\ell_2)}{(\text{Slope }\ell_3) - (\text{Slope }\ell_2)}\right)}{\left(\frac{(\text{Slope }\ell_1) - (\text{Slope }\ell_4)}{(\text{Slope }\ell_3) - (\text{Slope }\ell_4)}\right)} = -0.360$$

Figure 6.3.1 Computing the Cross Ratio

Let

- $\angle CPA = \tau_1$, $\angle CPB = \tau_2$, $\angle DPB = \tau_3$, and $\angle DPA = \tau_{123}$;
- Directed line segments are named using the ray AC as the positive direction: $d_1 = AC$, $d_2 = CB$, $d_3 = BD$, and $d_{123} = AD$;
- Line segments ℓ_1, ℓ_2, ℓ_3, and ℓ_4 are not directed and are all treated as positive numbers; and
- Angles measured in the counterclockwise direction are treated as positive.

Definition 6.3.1

For any four *concurrent* lines AP, BP, CP, and DP (CP within $\angle APB$ and BP within $\angle CPD$) the cross ratio $R(\angle CPA \angle CPB, \angle DPA \angle DPB)$ characterizes the partition of $\angle DPA$ by lines BP and CP, where

$$R(\angle CPA \;\angle CPB, \angle DPA \;\angle DPB) = \frac{\sin \angle CPA / -\sin \angle CPB}{\sin \angle DPA / \sin \angle DPB}$$

**Definition
6.3.2**

For any pencil of lines ℓ_1, ℓ_2, ℓ_3, and ℓ_4 and associated angles τ_1, τ_2, τ_3, and τ_{123}, the cross ratio $R(\tau_1\tau_2, \tau_{123}\;\tau_3)$ characterizes the partition of τ_{123} by lines ℓ_2 and ℓ_3, where

$$R(\tau_1\tau_2, \tau_{123}\;\tau_3) = \frac{\sin \tau_1 / \sin \tau_2}{\sin \tau_{123} / \sin \tau_3}$$

Using the Law of Sines, we may write

$$\frac{d_1}{\sin \tau_1} = \frac{\ell_2}{\sin \alpha_1}; \frac{d_2}{\sin \tau_2} = \frac{\ell_2}{\sin \alpha_3}; \frac{d_3}{\sin \tau_3} = \frac{\ell_3}{\sin(180 - \alpha_3)}; \frac{d_{123}}{\sin \tau_{123}} = \frac{\ell_4}{\sin \alpha_1}$$

Substitutions based on these relationships in the expression for $R(\tau_1\tau_2, \tau_{123}\;\tau_3)$ yields

$$R(\tau_1\tau_2, \tau_{123}\;\tau_3) = \frac{\sin \tau_1 / -\sin \tau_2}{\sin \tau_{123} / \sin \tau_3} = \frac{d_1 \sin \alpha_1 / \ell_2 / -d_2 \sin \alpha_3 / \ell_2}{d_{123} \sin \alpha_1 / \ell_4 / d_3 \sin(180 - \alpha_3) / \ell_4} = -\frac{d_1 / d_2}{d_{123} / d_3}$$

**Definition
6.3.3**

For any four *collinear points* A, B, C, and D (C between A and B), the cross ratio $R(AB, CD)$ characterizes the partition of segment AD by points B and C, where

$$R(AB, CD) = \frac{-CA/CB}{-DA/-DB}$$

In terms of Figure 6.3.1,

$$R(AB, CD) = \frac{\text{distances from } C \text{ to } A \text{ and } B}{\text{distances from } D \text{ to } A \text{ and } B} = \frac{-CA/CB}{-DA/-DB} = -\frac{d_1/d_2}{d_{123}/d_3}$$

This result shows that for a given pencil of lines ℓ_1, ℓ_2, ℓ_3, and ℓ_4 and associated angles τ_1, τ_2, τ_3, and τ_{123}, the partition of the field of view is equivalent to the cross ratio of the segments determined by the pencil of points $ACBD$.

**Theorem
6.3.1**

For a given set of concurrent lines (AP, CP, BP, and DP) and intercepted segments (CA, CB, DA, and DB), the cross ratio of angles $R(\angle CPA \;\angle CPB, \angle DPA \;\angle DPB)$ is equal to the cross ratio of intercepted segments $R(AB, CD)$.

Theorem 6.3.1 may be restated in a more powerful form.

**Theorem
6.3.2**

Given a perspectivity between a pencil of lines and a pencil of points, the cross ratio of the pencil of lines is equal to the cross ratio of the pencil of points.

If a different line m' intersects lines ℓ_1, ℓ_2, ℓ_3, and ℓ_4 in different points A', C', B' and D', a different set of intercepted segments d_1', d_2', and d_3' is obtained (see Figure 6.3.2). However, because the angles

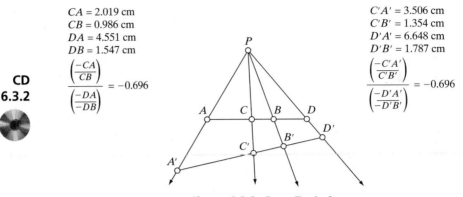

CD
6.3.2

Figure 6.3.2 Cross Ratio 2

formed by the concurrent lines remain unchanged, $R(\angle CPA \angle CPB,$ $\angle DPA \angle DPB) = R(\angle C'PA' \angle C'PB', \angle D'PA' \angle D'PB')$. By Theorem 6.3.1, this forces $R(AB, CD) = R(A'B', C'D')$. So the partition of the field of view may be determined by any set of intercepted segments.

Theorem 6.3.3

For a given set of concurrent lines $(AP, CP, BP, \text{ and } DP)$ the partition of the field of view from point P may be determined by any set of intercepted segments.

Theorem 6.3.3 also may be restated in a more powerful form.

Theorem 6.3.4

The cross ratio of a given pencil of lines is equal to the cross ratio of any pencil of points related to the pencil of lines by a perspectivity.

Figure 6.3.1 also contains a set of calculations based on the slopes of lines $\ell_1, \ell_2, \ell_3,$ and ℓ_4. From analytic geometry, we know that two lines with slopes u and v intersect in an angle τ given by

$$\tan \tau = \frac{u - v}{1 + uv}$$

If we let $a = $ slope of ℓ_1, $b = $ slope of ℓ_2, $c = $ slope of ℓ_3, and $d = $ slope of ℓ_4, we may write

$$\tan \tau_1 = \frac{a - c}{1 + ac}; \tan \tau_2 = \frac{b - c}{1 + bc}; \tan \tau_3 = \frac{b - d}{1 + bd}; \quad \text{and}$$

$$\tan \tau_{123} = \frac{a - d}{1 + ad}$$

In Figure 6.3.3, the first of these expressions is modeled in a right triangle. Using the Pythagorean Theorem, the hypotenuse $h = \sqrt{(a^2 + 1)(c^2 + 1)}$. We may then write

$$\sin \tau_1 = \frac{a - c}{\sqrt{(a^2 + 1)(c^2 + 1)}} \qquad \sin \tau_2 = \frac{b - c}{\sqrt{(b^2 + 1)(c^2 + 1)}}$$

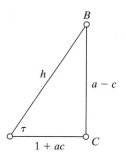

Figure 6.3.3 Cross Ratio by Slopes 1

$$\sin \tau_3 = \frac{b - d}{\sqrt{(b^2 + 1)(d^2 + 1)}} \qquad \sin \tau_{123} = \frac{a - d}{\sqrt{(a^2 + 1)(d^2 + 1)}}$$

Using the same approach taken in writing the cross ratio $R(\angle CPA \angle CPB,$ $\angle DPA \angle DPB)$ and simplifying terms, we obtain

$$\frac{\sin \angle CPA/-\sin \angle CPB}{\sin \angle DPA/\sin \angle DPB} = \frac{(a - c)/(b - c)}{(a - d)/(b - d)}$$

Theorem 6.3.5 For a given set of concurrent lines $(AP, CP, BP,$ and $DP)$ with slopes $a,$ $c, b,$ and $d,$ the cross ratio is given by

$$R(ac, bd) = \frac{(a - c)/(b - c)}{(a - d)/(b - d)}$$

Alternatively, we may write

Theorem 6.3.6 For a given pencil of lines with slopes $a, c, b,$ and $d,$ the cross ratio is given by

$$R(ac, bd) = \frac{(a - c)/(b - c)}{(a - d)/(b - d)}$$

Figure 6.3.4 offers insights into a related problem, characterizing the partition of a line using different sets of rays. Since both sets of rays intersect the given line in points $A, C, B,$ and $D,$ $R(\tau_1\tau_2, \tau_{123} \tau_3) = R(AB, CD) = R(\rho_1\rho_2, \rho_{123} \rho_3)$. In other words, when the same set of collinear points is observed from two different perspectives, the field of view is divided the same way as determined by the cross ratio.

Theorem 6.3.7 The cross ratio of a given pencil of points is equal to the cross ratio of any pencil of lines related to the pencil of points by a perspectivity.

The consequence of these findings is one of projective geometry's most useful results.

Theorem 6.3.8 The cross ratio is invariant under perspective transformations.

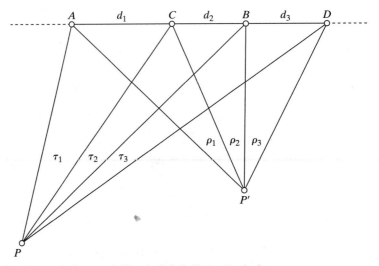

Figure 6.3.4 Cross Ratio 3

Harmonic Sets

Figure 6.3.5 shows a pencil of three points P, P_1, and P_2. The parameter $u = PP_1/PP_2$ is the algebraic ratio in which point P divides segment P_1P_2. Directed line segments PP_1 and PP_2 are arbitrarily assigned positive values. By contrast the directed segment P_1P would be assigned a negative value.

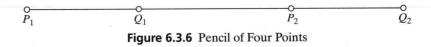

Figure 6.3.5 Directed Line Segments

Figure 6.3.6 shows a pencil of four points P_1, Q_1, P_2, Q_2. The parameter $u_1 = Q_1P_1/Q_1P_2$ is the algebraic ratio in which point Q_1 divides segment P_1P_2 and the parameter $u_2 = Q_2P_1/Q_2P_2$ is the algebraic ratio in which point Q_2 divides segment P_1P_2.

$$\overset{\circ}{P_1} \quad\quad \overset{\circ}{Q_1} \quad\quad\quad \overset{\circ}{P_2} \quad\quad\quad \overset{\circ}{Q_2}$$

Figure 6.3.6 Pencil of Four Points

**Definition
6.3.4**

Points Q_1 and Q_2 separate P_1 and P_2 *harmonically* if $u_1 = -u_2$.

**Definition
6.3.5**

If points Q_1 and Q_2 separate P_1 and P_2 harmonically, the points form a *harmonic set*.

Most students first encounter the term harmonic in a calculus course while studying sequences and series. Arithmetic sequences are characterized by an initial term and a constant difference between all subsequent terms. For instance, the sequence $2, 3, 4, 5 \ldots$ is arithmetic, having an initial term of 2 and a common difference of 1 between terms. The reciprocals of these terms, $1/2, 1/3, 1/4, 1/5, \ldots$, form a harmonic sequence. So, for every arithmetic sequence not including the number zero, there is an associated harmonic sequence, and vice versa. It is natural to wonder whether this use of the word harmonic is consistent with Definition 6.3.4. To answer this question, consider the differences between the reciprocals of successive terms.

Let $Q_1P_2 = d_1$. Then $Q_1P_1 = u_1d_1$. Let $Q_2P_2 = d_2$. Then $Q_2P_1 = u_2d_2 = -u_1d_2$. By writing the differences between reciprocals of successive terms appropriately, we can simplify $u_1 = -u_2 = u$. Since $ud_1 + d_1 + d_2 = ud_2$, we may then write $d_2 = d_1(u + 1)/(u - 1)$. Further simplification yields a first difference $[1/ud_1] - [1/(ud_1 + d_1)] = 1/u(u + 1)d_1$ and a second difference $[1/(ud_1 + d_1)] - [1/(ud_1 + d_1 + d_2)] = 1/u(u + 1)d_1$. So, the differences are the same and the two uses of harmonic are consistent.

CD 6.3.3

A remarkable relationship exists between harmonic sets and complete quadrangles. In Figure 6.3.7 a complete quadrangle is determined by points W, X, Y, Z and lines $a, b, c, d, e,$ and f. Points $P_1, Q_1, P_2,$ and Q_2 on line ℓ are determined by the complete quadrangle. The following argument demonstrates that these points form a harmonic set. Analytic geometry is used to simplify the demonstration.

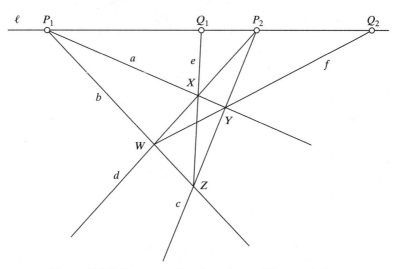

Figure 6.3.7 Complete Quadrangles and Harmonic Sets

Let P_1 correspond to $(0, 0)$ with P_2 to the right on the x-axis. The notation P_{2x} will be used to designate the x-coordinate of point P_2. Let lines $a, b, c,$ and d have slopes $m_a, m_b, m_c,$ and m_d respectively. Then the equations of lines $a, b, c,$ and d are $y = m_ax, y = m_bx, y = m_c(x - P_{2x}),$ and $y = m_d(x - P_{2x})$. These lines intersect at points

$$W = \left(\frac{-m_dP_{2x}}{m_b - m_d}, \frac{-m_bm_dP_{2x}}{m_b - m_d} \right), X = \left(\frac{-m_dP_{2x}}{m_a - m_d}, \frac{-m_am_dP_{2x}}{m_a - m_d} \right),$$

$$Y = \left(\frac{-m_cP_{2x}}{m_a - m_c}, \frac{-m_am_cP_{2x}}{m_a - m_c} \right), \text{ and } Z = \left(\frac{-m_cP_{2x}}{m_b - m_c}, \frac{-m_bm_cP_{2x}}{m_b - m_c} \right).$$

Lines e and f intersect the x-axis at points Q_1 and Q_2. The notations Q_{1x} and Q_{2x} will be used to designate the x-coordinate of points Q_1 and Q_2, respectively. If we let $\alpha = m_a m_b (m_c - m_d)/m_c m_d (m_a - m_b)$, the x-coordinate of Q_1 is given by $P_{2x}/(1 + \alpha)$ and the x-coordinate of Q_2 is given by $P_{2x}/(1 - \alpha)$.

To show that the segment lengths P_1Q_1, P_1P_2, and P_1Q_2 are in harmonic progression, we must show that the reciprocals of these lengths are in arithmetic progression. This is done by showing that there is a common difference between successive reciprocal terms. Since $d(P_1P_2) = P_{2x}$ and $d(P_1Q_1) = P_{2x}/(1 + \alpha)$, $1/P_{2x} - (1 + \alpha)/P_{2x} = -\alpha/P_{2x}$. This is the difference between the first two reciprocal terms. Next, since $d(P_1Q_2) = P_{2x}/(1 - \alpha)$, $(1 - \alpha)/P_{2x} - 1/P_{2x} = -\alpha/P_{2x}$. This is the difference between the second and third reciprocal terms. Because there is a common difference between successive reciprocal terms, the sequence of reciprocal terms is arithmetic. As a result, the sequence of segment lengths is harmonic. This result is stated formally in the following theorem.

Theorem 6.3.9

Four collinear points P_1, P_2, Q_1, Q_2 form a harmonic set $H(P_1P_2, Q_1Q_2)$ if there is a complete quadrangle in which one pair of opposite sides intersect at P_1, a second pair intersects at P_2, one diagonal line contains Q_1, and the other contains Q_2. Point Q_2 is called the harmonic conjugate of Q_1 with respect to points P_1 and P_2 (and vice versa).

Theorem 6.3.10

The cross ratio of a harmonic set is -1.

By Definitions 6.3.3 and 6.3.4, the cross ratio of a harmonic set may be computed as $u_1/u_2 = -1$.

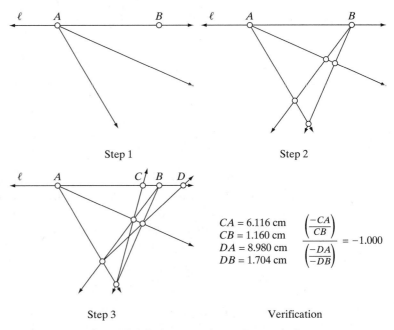

Step 1 Step 2

$CA = 6.116$ cm
$CB = 1.160$ cm
$DA = 8.980$ cm
$DB = 1.704$ cm

$$\dfrac{\left(\dfrac{-CA}{CB}\right)}{\left(\dfrac{-DA}{-DB}\right)} = -1.000$$

Step 3 Verification

Figure 6.3.8 Constructing a Harmonic Set

Figure 6.3.8 illustrates a method for constructing a harmonic set of points. The final frame in the figure verifies that the cross ratio determined by the segments is -1. The steps in the construction are

1. Select two points A and B on line ℓ. Draw two rays from point A as shown.
2. Draw a second pair of rays from point B so that they intersect the first two rays.
3. Draw the diagonal lines and find their intersections with line ℓ. Label the intersections C and D.

Summary

Perspectivities are the simplest projective transformations, relating pencils of points and lines in a direct way. The cross ratio provides a means for exploring the manner in which perspectivities divide the space that they occupy. Furthermore, the cross ratio demonstrates the existence of invariant properties under projective transformations. From these observations, many useful applications have emerged, making projective geometry a powerful tool in mathematics, science, and engineering. A few of these applications are explored in subsequent sections.

Cross Ratio Definitions
Q: For any four concurrent lines AP, BP, CP, and DP (CP within $\angle APB$ and BP within $\angle CPD$), how is the cross ratio computed?
A: The cross ratio $R(\angle CPA \angle CPB, \angle DPA, \angle DPB)$ is computed as

$$R(\angle CPA \angle CPB, \angle DPA \angle DPB) = \frac{\sin \angle CPA/-\sin \angle CPB}{\sin \angle DPA/\sin \angle DPB}$$

Q: For any four *collinear points* A, B, C, and D (C between A and B), how is the cross ratio computed?
A: The cross ratio $R(AB, CD)$ is computed as

$$R(AB, CD) = \frac{-CA/CB}{-DA/-DB}$$

Q: For a given set of concurrent lines $AP, CP, BP,$ and DP with slopes $a, c, b,$ and d, how is the cross ratio computed?
A: The cross ratio is computed as

$$R(ac, bd) = \frac{(a - c)/(b - c)}{(a - d)/(b - d)}$$

Cross Ratio Equivalence
Q: What relationship exists between the different definitions of the cross ratio and their computed values?
A: The definitions are logically equivalent and the computed cross ratio is the same whether approached using segments, angles, or slopes.

Harmonic Sets

Q: Given three collinear points P, P_1, and P_2, what algebraic relationship characterizes their relative positions on the line?

A: $u = PP_1/PP_2$ is the algebraic ratio in which point P divides segment P_1P_2.

Q: Under what conditions do four collinear points form a harmonic set?

A: When four collinear points P_1, P_2, Q_1, Q_2 form a harmonic set their cross ratio is -1.

Exercises

1. Line m is parallel to line n. Points A, B, C, and D on line m are projected through point P onto points A', B', C', and D' on line n. Using only Euclidean arguments, prove that the cross ratio of intercepted segments is the same for both sets of points.

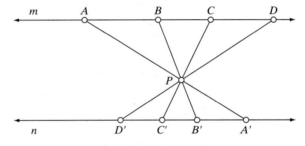

2. Construct the harmonic conjugate of C with respect to A and B.

3. Using the principal of duality, write the dual of Definition 6.3.5 and illustrate the new definition with a sketch.

4. Compute the cross ratios of the following pencil of points. Is the set harmonic?

5. Compute the cross ratios of the following pencil of lines. Is the set harmonic?

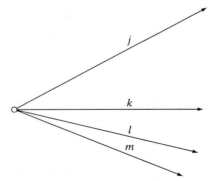

Investigations

Complete Quadrilaterals and Harmonic Sets

Tool(s) *Geometers Sketchpad*   *Data File(s)* cd6_3_3.gsp

Focus
Use a complete quadrilateral to divide a portion of a line harmonically.

Tasks
1. What happens to the relative positions of points Q_1 and Q_2 when you move
   a) Line *a*
   b) Line *b*
   c) Line *c*
   d) Line *d*
2. Measure the following distances:
   a) Q_1–P_1
   b) Q_1–P_2
   c) Q_2–P_1
   d) Q_2–P_2
3. Use the Calculate option under Measure to compute the following ratios:
   a) Q_1–P_1/Q_1–P_2
   b) Q_2–P_1/Q_2–P_2
   c) $(Q_1$–P_1/Q_1–$P_2)/(Q_2$–P_1/Q_2–$P_2)$

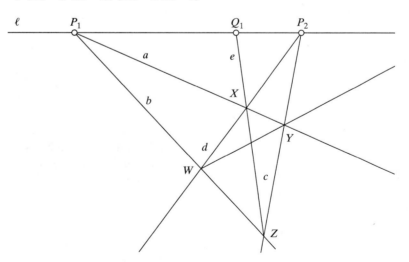

Dividing Space 1

Tool(s) *Geometers Sketchpad*   *Data File(s)* cd6_3_4.gsp

Focus
The cross ratio may be used as a indicator of how an object divides space. In this activity, the cross ratio is used to examine Alberti's Method.

Tasks

1. Describe the three sets of points for which the cross ratio is computed and their relationships to one another.
2. Vary the position of the Distance Line. Vary the position of the Horizon Line. Describe the effect of these actions on the cross ratios.
3. Do these observations strengthen Alberti's assertion that his method creates a true view of the world? If so, how? If not, why not?

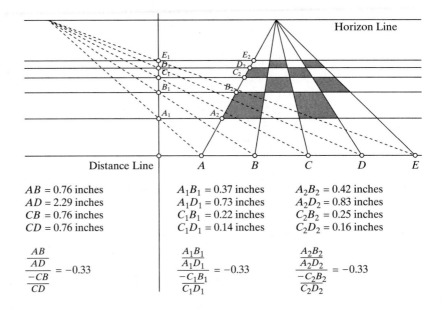

$AB = 0.76$ inches
$AD = 2.29$ inches
$CB = 0.76$ inches
$CD = 0.76$ inches

$$\frac{\frac{AB}{AD}}{\frac{-CB}{CD}} = -0.33$$

$A_1B_1 = 0.37$ inches
$A_1D_1 = 0.73$ inches
$C_1B_1 = 0.22$ inches
$C_1D_1 = 0.14$ inches

$$\frac{\frac{A_1B_1}{A_1D_1}}{\frac{-C_1B_1}{C_1D_1}} = -0.33$$

$A_2B_2 = 0.42$ inches
$A_2D_2 = 0.83$ inches
$C_2B_2 = 0.25$ inches
$C_2D_2 = 0.16$ inches

$$\frac{\frac{A_2B_2}{A_2D_2}}{\frac{-C_2B_2}{C_2D_2}} = -0.33$$

Dividing Space 2
Tool(s) Geometers Sketchpad   *Data File(s)* cd6_3_5.gsp

Focus
The cross ratio may be computed on the basis of either angular or linear measurements and used as an indicator of how an object divides the field of view of an observer. In cd6_3_5.gsp, the observer is at point P and the object is square $ABCD$.

Tasks

1. Describe the effect on the cross ratio of moving the line; resizing the square using the Side of Square segment; rotating the square using the Square drag point.
2. Vary the position of point P. Describe the effect on the cross ratio.
3. Do both methods of computing the cross ratio always return the same result?

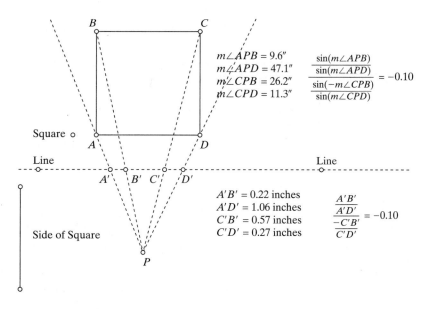

$m\angle APB = 9.6''$
$m\angle APD = 47.1''$
$m\angle CPB = 26.2''$
$m\angle CPD = 11.3''$

$$\frac{\frac{\sin(m\angle APB)}{\sin(m\angle APD)}}{\frac{\sin(-m\angle CPB)}{\sin(m\angle CPD)}} = -0.10$$

$A'B' = 0.22$ inches
$A'D' = 1.06$ inches
$C'B' = 0.57$ inches
$C'D' = 0.27$ inches

$$\frac{\frac{A'B'}{A'D'}}{\frac{-C'B'}{C'D'}} = -0.10$$

Dividing Space 3
Tool(s) Geometers Sketchpad   *Data File(s)* cd6_3_6.gsp

Focus
A comparison of the cross ratios of different quadrilaterals

Tasks
1. Reposition points A, B, C, and D, creating a different quadrilateral. What happens to the cross ratio?
2. Is it possible for different quadrilaterals to have the same cross ratio relative to some observer? If so, under what circumstances?
3. Discuss the implications of your observations for the following situation: You are standing at point P on a large plot of land, drawing a map showing the location of buildings and other easily recognized objects.

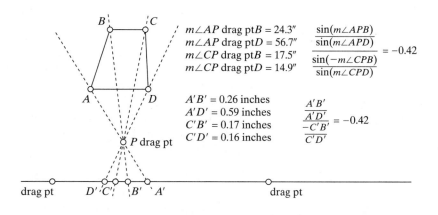

$m\angle AP$ drag pt$B = 24.3''$
$m\angle AP$ drag pt$D = 56.7''$
$m\angle CP$ drag pt$B = 17.5''$
$m\angle CP$ drag pt$D = 14.9''$

$$\frac{\frac{\sin(m\angle APB)}{\sin(m\angle APD)}}{\frac{\sin(-m\angle CPB)}{\sin(m\angle CPD)}} = -0.42$$

$A'B' = 0.26$ inches
$A'D' = 0.59$ inches
$C'B' = 0.17$ inches
$C'D' = 0.16$ inches

$$\frac{\frac{A'B'}{A'D'}}{\frac{-C'B'}{C'D'}} = -0.42$$

6.4 Applications of the Cross Ratio

In this section,
you will . . .

Investigate applications of the cross ratio in
• **Surveying;**
• **Art;**
• **Remote Sensing.**

The examples in this section illustrate the use of the cross ratio in a variety of contexts.

**Example
6.4.1**

An observer positioned as shown in Figure 6.4.1 wants to estimate the distance across the water from point B to point D, labeled δ_3. The following information is known: $\tau_1 = 15°$, $\tau_2 = 10°$, $\tau_3 = 20°$, $\delta_1 = 200$ m, $\delta_2 = 100$ m.

By Theorem 6.3.6, we know that the pencil of lines and the pencil of points in the figure have the same cross ratio. We may represent this in the equation

$$\frac{\sin 15/\sin 10}{\sin 45/\sin 20} = \frac{200/100}{(300 + \delta_3)/\delta_3}$$

Solving for δ_3 yields the value 169 m.

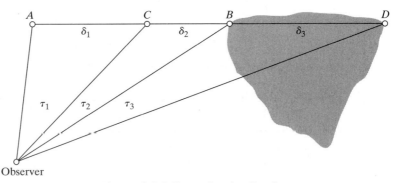

Figure 6.4.1 Surveying Application

**Example
6.4.2**

An illustrator wants to draw a set of railroad tracks converging in the distance. As shown in Figure 6.4.2, the distance between the first three ties in the drawing is known: 1 inch between the first and second ties, 9/11 of an inch between the second and third tracks. The illustrator must determine how to space the remaining ties and where to position the vanishing point. Because the two images represent different perspective views of the same set of lines, they have the same cross ratio. As a result, we may write the equation

$$\frac{-CA/CB}{-D_nA/-D_nB} = \frac{-RP/RQ}{-S_nP/-S_nQ}$$

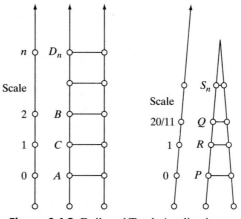

Figure 6.4.2 Railroad Track Application

Substituting the known distances in this equation yields

$$\frac{-1/1}{-n/-(n-2)} = \frac{-1/(9/11)}{-S_n/-(S_n - 20/11)}$$

Solving for S_n yields the formula

$$S_n = \frac{10n}{n+9}$$

The position of any railroad tie may be computed by substituting values for n:

$$S_0 = \frac{10(0)}{(0)+9} = 0; \quad S_1 = \frac{10(1)}{(1)+9} = 1; \quad S_2 = \frac{10(2)}{(2)+9} = 20/11; \quad S_3 = \frac{10(3)}{(3)+9} = 30/12;$$

and so on

The position of the vanishing point may be found by computing the limit as $n \to \infty$.

$$\lim_{n\to\infty} \frac{10n}{n+9} = \lim_{n\to\infty} \frac{10}{1+9/n} = 10$$

Example 6.4.3

URL 6.4.1

Figures 6.4.4 and 6.4.5 are NASA satellite images of the same canyon system on Mars, seen from the south and from the west. Five points are labeled in each figure, $A, B, C, D,$ and E. Point X appears in Figure 6.4.4 but not in Figure 6.4.5. To estimate the position of point X in Figure 6.4.4, it is easiest to use an old map maker's trick called the *paper strip technique*. The basis for the paper strip technique is illustrated in Figure 6.4.3.

In a typical application of the paper strip technique, a photograph (Photo Plane View) shows the location of points $A, B, C, D,$ and E. A perspective view of this sort might be obtained from a mountain top, an aircraft, or a satellite. The Map Plane View contains associated points

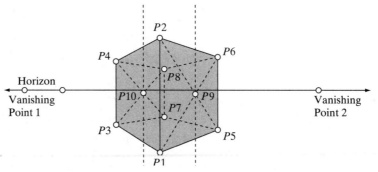

Figure 6.4.3 Paper Strip Technique

A, B, C, and D, but not E. The strategy for locating the position of point E in the Map Plane View involves finding two lines, both of which are known to contain the point E. Point E is located at the intersection of those lines. The steps implementing this strategy are as follows:

1. As seen in Figure 6.4.3, a pencil of lines is drawn from point B to points A, C, D, and E in the Photo Plane View. A corresponding pencil of lines is drawn from point B to points A, C, and D in the Map Plane View.
2. The edge of a strip of paper is laid across the pencil of lines in the Photo Plane View. The edge of the paper strip is marked with intersection points A', C', D', and E'.
3. The strip is then moved to the Map Plane View where the strip is positioned so that points A', C', and D' on the edge of the strip line up with lines BA, BC, and BD respectively.
4. The point in the Map Plane View next to point E' on the paper strip is marked and a line is drawn from B through that point. This is the first of the two lines needed to find E in the Map Plane View.
5. Steps 1–4 are repeated using point A as the center of the pencil of lines. This results in a second line known to contain point E.
6. Point E is located at the intersection of these two lines.

Applying the paper strip technique to the task of locating the position of point X in Figures 6.4.6 – 6.4.7 is approached in the same manner. It should be noted that the paper strip technique and the cross ratio itself

CD
6.4.1

Figure 6.4.4 View North, Martian Canyon

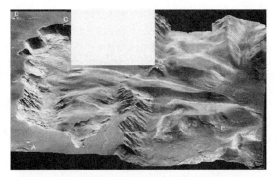

Figure 6.4.5 View East, Martian Canyon

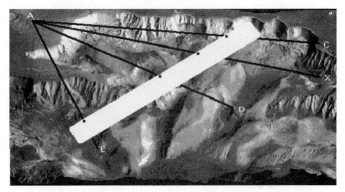

Figure 6.4.6 Paper Strip 1

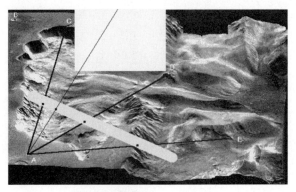

Figure 6.4.7 Paper Strip 2

are based on all points being coplanar. In the case of the Martian surface, this is clearly not the case. As a result, the position obtained for point X will be inaccurate. In a more rigorous approach, errors of this sort would be minimized through the use of more elaborate mathematical models.

Summary

Each application in this section makes use of the fact that the cross ratio is preserved under projective transformations and is the same whether computed using segments, angles, or slopes. The paper strip technique captures these relationships in a purely mechanical procedure. This remarkable technique has been used by generations of surveyors, cartographers, and students.

URLs	Note: Begin each URL with the prefix http://
6.4.1	d www.msss.com/newhome.html

Table 6.4.1 Section 6.4 URL (c = concept, h = history, s = software, d = data)

Exercises

1. Under what circumstances is betweenness of points preserved under projective transformations?
2. Use the cross ratio to locate the position of the next three railroad ties and the vanishing point in the perspective view below.

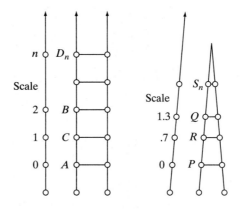

3. In an aerial photograph of a straight highway, the distances between points A and B, B and C, C and D are $1''$, $2''$ and $1''$, respectively. From older maps, it is known that the actual distances between points A' and B' and points B' and C' is 3 km and 2 km, respectively. Use the cross ratio to determine the actual distance between points C' and D'.

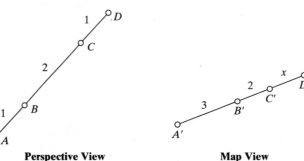

Perspective View   **Map View**

Investigation

Paper Strip Technique
Tool(s) Geometers Sketchpad   *Data File(s)* cd6_4_2.gsp

Focus

The paper strip technique is a mapmaker's trick for converting survey data into map data. In the Photo Plane View shown below, points $A, B, C, D,$ and E are marked. In the Map Plane View, only points $A, B, C,$ and D are seen.

Task

The task is to use the data in the Photo Plane Image to construct the location of point E in the Map Plane Image.

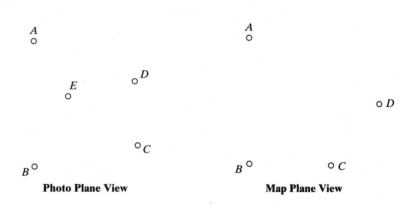

Photo Plane View   Map Plane View

6.5 Matrix Methods for 3-Point Perspective Transformations

In this section, you will . . .

Use matrix methods to:
- **Represent points in 3-space;**
- **Compute the position of points in 3-space under perspective transformations;**
- **Compute the position of vanishing points associated with sets of parallel lines.**

URLs
6.5.1
&
6.5.2

Most computer applications of projective geometry use matrix representations and operations to determine the positions of points in various perspective views. These representations and operations are normally not displayed in the user interfaces, since most people are not concerned with the underlying mathematics. This section focuses on the power and elegance of the mathematics itself.

Transformation Matrices

Figure 6.5.1 shows a unit cube located in the corner of the first octant. The view is not perspective in nature, since none of the parallel lines converge to vanishing points. Imagine yourself positioned on the z-axis at a point z_c located several units from the origin. If the cube is solid, only one face would be visible from z_c. For most purposes, perspective views of this sort are neither interesting nor useful.

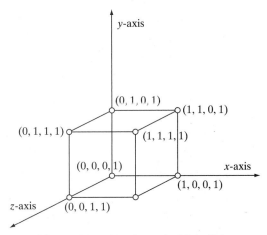

Figure 6.5.1 Vertices of a Unit Cube

One strategy for improving the perspective is to rotate the cube around one or more axes, bringing more faces into view. For instance, the cube could be rotated first about the y-axis through an angle of $-30°$ (clockwise, exposing the face on the right side of the cube), then about the *x*-axis through an angle of 45° (counterclockwise, exposing the top face of the cube). Using a perspective transformation, the cube is then projected onto the *x-y* plane for viewing. The actual steps are

1. Represent the cube's vertices in matrix form.
2. Represent both rotations in matrix form
   $\phi°$ about the *y*-axis;
   $\theta°$ about the *x*-axis.
3. Represent a projection of the vertices onto the *x-y* plane in matrix form.
4. Compose the transformation matrices.
5. Find the image of the vertices under the composition of matrices.
6. Connect the vertices to indicate the edges and faces of the cube.

Step 1: Since 3-dimensional objects have *x*, *y*, and *z* coordinates, matrix representations of points in 3-space are of the form $(x, y, z, 1)$. Using this representation, the eight vertices of the cube may be consolidated into a single matrix

$$X = \begin{bmatrix} 0 & 1 & 1 & 0 & 0 & 1 & 1 & 0 \\ 0 & 0 & 1 & 1 & 0 & 0 & 1 & 1 \\ 1 & 1 & 1 & 1 & 0 & 0 & 0 & 0 \\ 1 & 1 & 1 & 1 & 1 & 1 & 1 & 1 \end{bmatrix}$$

Step 2: Matrix representations for Euclidean transformations in 3-space are similar in form to the matrices presented in Chapter Four. For instance, the matrix representation for a rotation about the *x*-axis is given by

$$R_x = \begin{bmatrix} 1 & 0 & 0 & 0 \\ 0 & \cos\theta & -\sin\theta & 0 \\ 0 & \sin\theta & \cos\theta & 0 \\ 0 & 0 & 0 & 1 \end{bmatrix}$$

The image of a point $(x, y, z, 1)$ under this transformation is given by

$$\begin{bmatrix} 1 & 0 & 0 & 0 \\ 0 & \cos\theta & -\sin\theta & 0 \\ 0 & \sin\theta & \cos\theta & 0 \\ 0 & 0 & 0 & 1 \end{bmatrix}\begin{bmatrix} x \\ y \\ z \\ 1 \end{bmatrix} = \begin{bmatrix} x \\ y\cos\theta - z\sin\theta \\ y\sin\theta + z\cos\theta \\ 1 \end{bmatrix}$$

Notice that under a rotation about the *x*-axis, the *x*-coordinate of a point is fixed and its *y* and *z* coordinates change.

The matrix representation of a rotation about the *y*-axis is given by

$$R_y = \begin{bmatrix} \cos\phi & 0 & \sin\phi & 0 \\ 0 & 1 & 0 & 0 \\ -\sin\phi & 0 & \cos\phi & 0 \\ 0 & 0 & 0 & 1 \end{bmatrix}$$

The image of the point $(x, y, z, 1)$ under this transformation is given by

$$\begin{bmatrix} \cos \phi & 0 & \sin \phi & 0 \\ 0 & 1 & 0 & 0 \\ -\sin \phi & 0 & \cos \phi & 0 \\ 0 & 0 & 0 & 1 \end{bmatrix} \begin{bmatrix} x \\ y \\ z \\ 1 \end{bmatrix} = \begin{bmatrix} x \cos \phi + z \sin \phi \\ y \\ -x \sin \phi + z \cos \phi \\ 1 \end{bmatrix}$$

Under rotations about the y-axis, the y-coordinate of a point remains fixed and its x and z coordinates change. Notice that the signs on the $\sin \phi$ terms are different than in the matrix R_x. These differences are a result of the way that positive angles are defined in a standard right hand coordinate system.

Step 3: Figure 6.5.2 shows a "side view" of the x-y plane. A line connects the center of the perspectivity z_c through vertex P to its image in the x-y plane P'.

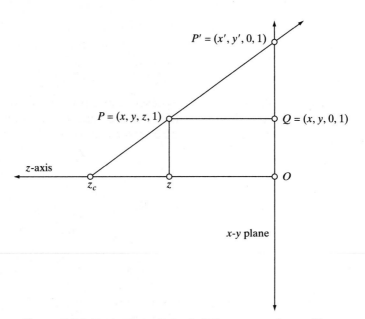

Figure 6.5.2 Projecting a Point in 3-Space onto the x-y Plane

Using similar triangles,

$$\frac{z_c - z}{OQ} = \frac{z_c}{OP'} \Rightarrow OP' = OQ\left(\frac{z_c}{z_c - z}\right), \quad \text{where} \quad OQ = \sqrt{x^2 + y^2} \quad \text{and}$$

$$OP' = \sqrt{x'^2 + y'^2}$$

Since the same factor used to rescale OQ to OP' applies to the x and y coordinates,

$$P' = \begin{bmatrix} x' \\ y' \\ 0 \\ 1 \end{bmatrix} = \begin{bmatrix} x\left(\dfrac{z_c}{z_c - z}\right) \\ y\left(\dfrac{z_c}{z_c - z}\right) \\ 0 \\ 1 \end{bmatrix}$$

This same result is obtained using the matrix P_z, where

$$P_z = \begin{bmatrix} 1 & 0 & 0 & 0 \\ 0 & 1 & 0 & 0 \\ 0 & 0 & 0 & 0 \\ 0 & 0 & -\dfrac{1}{z_c} & 1 \end{bmatrix}$$

$$P_z X = \begin{bmatrix} 1 & 0 & 0 & 0 \\ 0 & 1 & 0 & 0 \\ 0 & 0 & 0 & 0 \\ 0 & 0 & -\dfrac{1}{z_c} & 1 \end{bmatrix} \begin{bmatrix} x \\ y \\ z \\ 1 \end{bmatrix} = \begin{bmatrix} x \\ y \\ 0 \\ \left(-\dfrac{z}{z_c} + 1\right) \end{bmatrix} = \begin{bmatrix} x \\ y \\ 0 \\ \left(\dfrac{z_c - z}{z_c}\right) \end{bmatrix} = \begin{bmatrix} x\left(\dfrac{z_c}{z_c - z}\right) \\ y\left(\dfrac{z_c}{z_c - z}\right) \\ 0 \\ 1 \end{bmatrix}$$

Step 4: Composing these transformations yields the matrix

$$P_z R_x R_y = \begin{bmatrix} \cos\phi & 0 & \sin\phi & 0 \\ \sin\phi\sin\theta & \cos\theta & -\cos\phi\sin\theta & 0 \\ 0 & 0 & 0 & 0 \\ \dfrac{\sin\phi\cos\theta}{z_c} & \dfrac{-\sin\theta}{z_c} & \dfrac{-\cos\phi\cos\theta}{z_c} & 1 \end{bmatrix}$$

Step 5: Letting $\phi = -30°$, $\theta = 45°$, and $z_c = 2.5$

$$X' = P_z R_x R_y X$$

$$= \begin{bmatrix} .866 & 0 & -.5 & 0 \\ -.354 & .707 & -.612 & 0 \\ 0 & 0 & 0 & 0 \\ -.141 & -.283 & -.245 & 1 \end{bmatrix} \begin{bmatrix} 0 & 1 & 1 & 0 \\ 0 & 0 & 1 & 1 \\ 1 & 1 & 1 & 1 \\ 1 & 1 & 1 & 1 \end{bmatrix} \begin{bmatrix} 0 & 1 & 1 & 0 \\ 0 & 0 & 1 & 1 \\ 0 & 0 & 0 & 0 \\ 1 & 1 & 1 & 1 \end{bmatrix}$$

$$= \begin{bmatrix} -.500 & .366 & .366 & -.500 & 0 & .866 & .866 & 0 \\ -.612 & -.966 & -.259 & .095 & 0 & -.354 & .353 & .707 \\ 0 & 0 & 0 & 0 & 0 & 0 & 0 & 0 \\ .755 & .614 & .331 & .472 & 1 & .859 & .576 & .717 \end{bmatrix}$$

Since the matrix operation does not return points in the format $(x, y, z, 1)$, the elements of each column are divided by the last entry in each column to obtain the proper format.

$$\begin{bmatrix} -.662 & .596 & 1.107 & -1.059 & 0 & 1.009 & 1.504 & 0 \\ -.811 & -1.574 & -.782 & .201 & 0 & -.412 & .614 & .986 \\ 0 & 0 & 0 & 0 & 0 & 0 & 0 & 0 \\ 1 & 1 & 1 & 1 & 1 & 1 & 1 & 1 \end{bmatrix}$$

Step 6: These points are plotted in Figure 6.5.3 and connected appropriately to produce the perspective image of the cube.

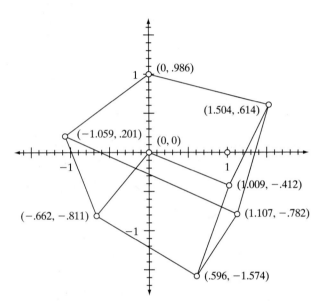

Figure 6.5.3 Perspective View of Cube

Vanishing Points

The perspective view of the cube seen in Figure 6.5.3 implies the existence of three vanishing points, each associated with lines parallel to one of the three coordinate axes. These vanishing points may be determined graphically by extending the edges of the cube until they intersect. The following matrix equation demonstrates an analytic approach to this task.

$$VP_{cube} = \begin{bmatrix} .866 & 0 & -.5 & 0 \\ -.354 & .707 & -.612 & 0 \\ 0 & 0 & 0 & 0 \\ -.141 & -.283 & -.245 & 1 \end{bmatrix} \begin{bmatrix} 1 & 0 & 0 \\ 0 & 1 & 0 \\ 0 & 0 & 1 \\ 0 & 0 & 0 \end{bmatrix} = \begin{bmatrix} -6.142 & 0 & 2.04 \\ 2.5 & -2.5 & 2.5 \\ 0 & 0 & 0 \\ 1 & 1 & 1 \end{bmatrix}$$

In this approach, the composite transformation matrix is multiplied by three points, each of which defines a vector from the origin in the direction of a set of parallel lines. For instance, a vector drawn from the origin to the point $(1, 0, 0, 0)$ is parallel to the four edges of the cube parallel to the *x*-axis. Similarly, vectors drawn from the origin to the points (0, 1, 0, 0) and (0, 0, 1, 0) are parallel to the edges of the cube that are parallel to the *y*- and *x*-axes, respectively. The points determined by this matrix equation are the coordinates of the projections of the vanishing points in the *x-y* plane.

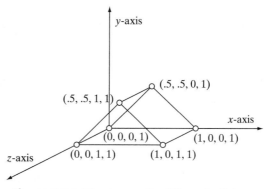

Figure 6.5.4 Nonperspective View of a Prism

Example 6.5.1

Figure 6.5.4 shows a prism in a nonprojective view. Letting $\phi = -30°$, $\theta = 45°$, and $z_c = 2.5$, the same procedures result in the perspective view seen in Figure 6.5.5. The matrix representations and operations are

$$X = \begin{bmatrix} 0 & 1 & .5 & 0 & 1 & .5 \\ 0 & 0 & .5 & 0 & 0 & .5 \\ 1 & 1 & 1 & 0 & 0 & 0 \\ 1 & 1 & 1 & 1 & 1 & 1 \end{bmatrix} \text{ and } P_z R_x R_y X = \begin{bmatrix} .866 & 0 & -.5 & 0 \\ -.354 & .707 & -.612 & 0 \\ 0 & 0 & 0 & 0 \\ -.141 & -.283 & -.245 & 1 \end{bmatrix}$$

$$\text{Then } X' = \begin{bmatrix} .866 & 0 & -.5 & 0 \\ -.354 & .707 & -.612 & 0 \\ 0 & 0 & 0 & 0 \\ -.141 & -.283 & -.245 & 1 \end{bmatrix} \begin{bmatrix} 0 & 1 & .5 & 0 & 1 & .5 \\ 0 & 0 & .5 & 0 & 0 & .5 \\ 1 & 1 & 1 & 0 & 0 & 0 \\ 1 & 1 & 1 & 1 & 1 & 1 \end{bmatrix}$$

$$= \begin{bmatrix} -.662 & .596 & -.123 & 0 & 1.009 & .55 \\ -.811 & -.1574 & -.802 & 0 & -.412 & .224 \\ 0 & 0 & 0 & 0 & 0 & 0 \\ 1 & 1 & 1 & 1 & 1 & 1 \end{bmatrix}$$

To determine the vanishing points associated with the parallel lines in Figure 6.5.4, a vector is defined from the origin in the direction of each set of parallel lines. These directions correspond to the positive *x*- and *z*-directions and the sloping "top" faces of the prism. The matrix representation of these vectors is

$$\begin{bmatrix} 1 & 0 & 1 & 1 \\ 0 & 0 & 1 & -1 \\ 0 & 1 & 0 & 0 \\ 0 & 0 & 0 & 0 \end{bmatrix}$$

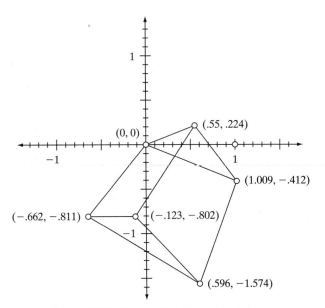

Figure 6.5.5 Perspective View of a Prism

The vanishing points are then computed using the matrix equation

$$
\mathbf{VP}_{prism} =
\begin{bmatrix}
.866 & 0 & -.5 & 0 \\
-.354 & .707 & -.612 & 0 \\
0 & 0 & 0 & 0 \\
-.141 & -.283 & -.245 & 1
\end{bmatrix}
\begin{bmatrix}
1 & 0 & 1 & 1 \\
0 & 0 & 1 & -1 \\
0 & 1 & 0 & 0 \\
0 & 0 & 0 & 0
\end{bmatrix}
$$

$$
=
\begin{bmatrix}
-6.14 & 2.04 & -2.04 & 6.09 \\
2.51 & 2.49 & -.83 & -7.47 \\
0 & 0 & 0 & 0 \\
1 & 1 & 1 & 1
\end{bmatrix}
$$

Summary

The procedures discussed in this section illustrate a computational approach to rendering realistic perspective views of 3-dimensional objects on 2-dimensional surfaces such as paper, computer screens, and movie screens. In the arts, science, and engineering, projective techniques such as these are making communication and entertainment more engaging and more convincing than ever. At the foundation of these efforts is projective geometry.

Points

Q: How are ordinary Euclidean points represented?

A: The Euclidean point (x, y, z) is written as a column vector

$$\begin{bmatrix} x \\ y \\ z \\ 1 \end{bmatrix}$$

Q: How are ideal (vanishing) points represented?

A: Ideal points are used to represent directions. The directions corresponding to the coordinate axes are written as

$$x\text{-axis} = \begin{bmatrix} 1 \\ 0 \\ 0 \\ 0 \end{bmatrix} \quad y\text{-axis} = \begin{bmatrix} 0 \\ 1 \\ 0 \\ 0 \end{bmatrix} \quad z\text{-axis} = \begin{bmatrix} 0 \\ 0 \\ 1 \\ 0 \end{bmatrix}$$

Rotations In 3-Space

Q: How is a rotation about the x-axis represented?

A: $R_x = \begin{bmatrix} 1 & 0 & 0 & 0 \\ 0 & \cos\theta & -\sin\theta & 0 \\ 0 & \sin\theta & \cos\theta & 0 \\ 0 & 0 & 0 & 1 \end{bmatrix}$

Q: How is a rotation about the y-axis represented?

A: $R_y = \begin{bmatrix} \cos\phi & 0 & \sin\phi & 0 \\ 0 & 1 & 0 & 0 \\ -\sin\phi & 0 & \cos\phi & 0 \\ 0 & 0 & 0 & 1 \end{bmatrix}$

Projection

Q: What transformation projects an object in 3-space onto the x-y plane from the point $(0, 0, z_c, 1)$ on the z-axis?

A: $P_z = \begin{bmatrix} 1 & 0 & 0 & 0 \\ 0 & 1 & 0 & 0 \\ 0 & 0 & 0 & 0 \\ 0 & 0 & -1/z_c & 1 \end{bmatrix}$

Composition of Transformations

Q: What transformation is equivalent to a composition of the following transformations: First about the *x*-axis, then about the *y*-axis, then projected onto the *x-y* plane.

$$
\text{A: } P_z R_x R_y =
\begin{bmatrix}
\cos\phi & 0 & \sin\phi & 0 \\
\sin\phi \sin\theta & \cos\theta & -\cos\phi \sin\theta & 0 \\
0 & 0 & 0 & 0 \\
\dfrac{\sin\phi \cos\theta}{z_c} & \dfrac{\sin\theta}{z_c} & \dfrac{-\cos\phi \cos\theta}{z_c} & 1
\end{bmatrix}
$$

URLs		Note: Begin each URL with the prefix http://
6.5.1	s	www.frontiernet.net/~imaging/java-3d-engine.html
6.5.2	s	www.nirim.go.jp/~weber/JAVA/jpoly/jpoly.html

Table 6.5.1 Section 6.5 URLs (c = concept, h = history, s = software, d = data)

Exercises

1. Write matrices representing the vertices of the following objects:

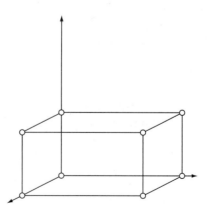

Right Square Prism Dimensions $1 \times 1 \times 2$

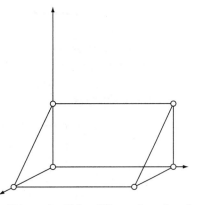

Right Triangular Prism Dimensions $1 \times 1 \times 2$

2. Write a transformation matrix for a Euclidean rotation in 3-space about the z-axis. Which parameter stays the same for all points and angles? Why?
3. Sketch the solid object with vertices given by the matrix

$$\begin{bmatrix} 0 & 0 & 1 & 1 & 0 & 0 & 1 & 1 \\ 0 & 1 & 0 & -1 & 0 & 1 & 0 & -1 \\ 0 & 0 & 0 & 0 & 1 & 1 & 1 & 1 \\ 1 & 1 & 1 & 1 & 1 & 1 & 1 & 1 \end{bmatrix}$$

4. Create a perspective view of the object in exercise 3, first rotating $30°$ about the x-axis, then rotating $-30°$ about the y-axis, then projecting the points onto the x-y plane using $z_c = 3$. Find all vanishing points.
5. Repeat the transformation in exercise 3, reversing the order of the rotations. Find all vanishing points. Do you expect to get the same result?

6.6 Applications of Geometry in Remote Sensing

In this section, you will . . .

- **Learn basic concepts and terminology related to remote sensing and image processing;**
- **Use the image processing program Scion Image and NASA data to measure the volcano Olympus Mons on the planet Mars.**

URLs 6.6.1 – 6.6.5

Introduction to Remote Sensing and Image Processing

Remote sensing is sensing from a distance. In most applications, remote sensing is used to acquire information for display in a visual format such as photographs and radar images. Remote Sensing extends our understanding of the world by using aircraft, spacecraft, and satellites to gather and interpret atmospheric, oceanographic, and terrestrial data. These data are used to predict the weather, understand phenomena such as El Niño, study agricultural problems, identify natural resources, aid cartographers in map making, and model local and global changes in the environment and climate.

URL 6.6.6

Image processing is used to manipulate image data in order to extract information not readily apparent on the basis of an informal visual inspection. Images may be measured to obtain lengths, perimeters, and areas, colored artificially to facilitate pattern recognition, combined to make comparisons easier, and so on.

Digital images are composed of arrays of numbers. In most images, these numbers are associated with specific colors or shades of gray. For instance, Table 6.6.1 shows an 8×8 array of integers. In Figure 6.6.1,

1	2	3	3	3	2	1	1
1	1	2	2	3	2	1	1
1	1	2	3	4	3	2	1
1	2	3	4	4	3	2	1
1	2	3	4	5	4	3	2
2	2	4	5	5	5	3	2
1	1	5	5	6	5	1	1
1	1	1	2	2	2	1	1

Table 6.6.1 Sample Array

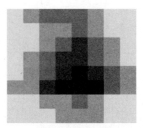

Figure 6.6.1 Sample Display

these integers are displayed as shades of gray. Each element in the array is associated with a single picture element, or pixel.

**URL
6.6.7**

Because the smallest unit of information in digital images is the pixel, no resolution of details at a finer scale is possible. Zooming in on a digital image eventually reveals its discrete structure. For instance, Figures 6.6.2–6.6.4 show zooms on an image of the volcano Olympus Mons on the planet Mars down to the level of individual pixels. In the images of Olympus Mons, each pixel represents a square on the ground approximately 1850 m on a side. This region is known as the *foot print* of the pixel and is the smallest area discernable in the image.

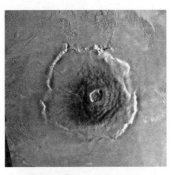

Figure 6.6.2 Zoom 1

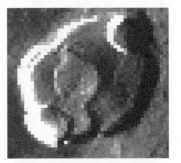

Figure 6.6.3 Zoom 2

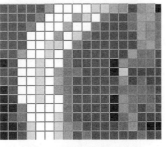

Figure 6.6.4 Zoom 3

The size of the foot print in a given digital image depends on two factors: The distance of the observer from the surface of interest (r) and the angular resolution of the camera (ρ). In Figure 6.6.5, the length of

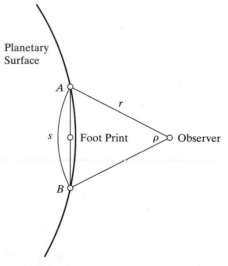

Figure 6.6.5 Pixel Foot Prints

arc s is given by $s = r\rho$, where r is measured in radians. For instance, on February 13, 1979 the video cameras on the Voyager spacecraft began taking pictures of Jupiter and its satellites. The Voyager optical system (crude by today's standards) had a resolution of about 10 microradians per pixel. As a result, this system could distinguish objects roughly 1 km (100,000 km *.000010 microradians) in diameter from a distance of approximately 100,000 km.

**CD
6.6.1**

As shown in Figure 6.6.6, foot prints (arc lengths) vary in size from one pixel to another when viewing a curved surface. This effect is diminished when the field of view spans only a small portion of the curved surface and when the position of the observer is a great distance from the surface. When the portion of the observed surface is large and the observer relatively close to the surface, pixels may vary in size significantly. Since measurements in most image processing systems assume a

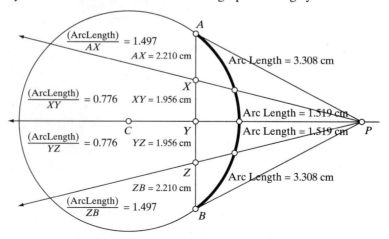

Figure 6.6.6 Foot Print Inconsistencies

consistent foot print, significant errors may occur when measuring length, perimeter, or area. This possibility should be kept in mind whenever performing measurements in an image processing system.

Measuring Olympus Mons

One of Mars' most spectacular features is the gigantic shield volcano called Olympus Mons. Located at 133 degrees West Longitude and 20 degrees North Latitude, Olympus Mons rises to an elevation of approximately 24,000 m, far above the thin Martian atmosphere. By contrast, Mt. Everest and the Hawaiian volcano Kilauea (measured from its base at the sea floor) are only about 9,000 m in height. The largest known volcano in the solar system, Olympus Mons is a true giant.

**CD
6.6.2**

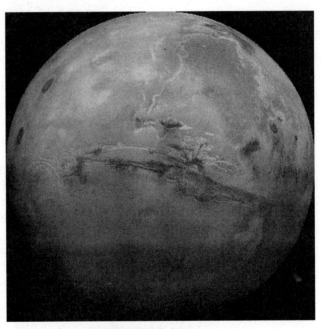

Figure 6.6.7 Mars

**URL
6.6.8**

Start *Scion Image*. When the program is loaded, move the pointer to the **File** pull-down menu in the upper left hand corner of the screen, depress the mouse button, drag down to **Open,** and release the button. Double-click on the file Mars.tif. After a few seconds an image will appear (see Figure 6.6.7). What is your first impression of the planet Mars? How would you describe Mars to someone who has never seen it? When you are finished, select **Close** from the File pull-down menu.

Open the file Mons. When the image appears, examine it carefully (see Figure 6.6.8). Select the magnifying glass in the upper left hand corner of the Tools window and move it onto the image. Click the mouse button repeatedly until you can see the individual pixels. Undo the magnification by double clicking on the magnifying glass in the Tools

CD
6.6.3

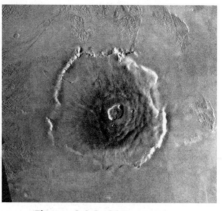

Figure 6.6.8 Olympus Mons

window. Each pixel in the Olympus Mons image is 1850 m on a side. How many kilometers is that? Suppose that a line on the image is 100 pixels long. How many kilometers is that?

Move the pointer to the Analyze pull-down menu at the top of the screen, depress the mouse button, drag down to Set Scale, and let go. A new window will open. In the center of the window is a box labeled Units. Click on the down arrow in the box and select kilometers as your unit of measurement. At the top of the window enter 1 in the Measured Distance box and 1.850 in the Known Distance box. Click OK once you have done so.

Using the Segment Tool (Tools window, 5th icon from the top in the right hand column), draw the longest segment you can across the entire volcano (see Figure 6.6.9). Once the segment is drawn, return to the Analyze pull-down menu, drag down to Measure, and release the mouse button. Again, click on Analyze, drag down to Show Results, and release the mouse button. A Results window will appear with the measurement in the column marked length. Take several measurements. Which is the "best" in your judgement? Why?

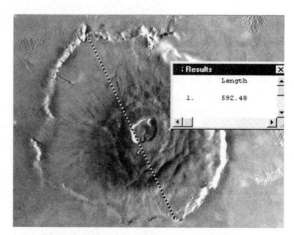

Figure 6.6.9 Measuring Olympus Mons

Use the magnifying glass in the Tools window to zoom in on the caldera (the crater in the center of the volcano). Take several measurements across the caldera using the Segment Tool. Which is the "best" in your judgement? Why? Using a calculator, determine the area of each pixel in square meters (recall that the image is made of pixels 1850 m on a side). Compute the area of each pixel in square kilometers. Is each pixel more than or less than one square kilometer?

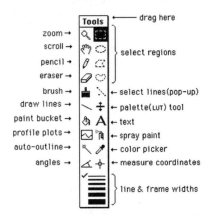

Figure 6.6.10 *Scion Image* Selection Tools

Scion Image has 4 tools for selecting regions (see Figure 6.6.10). The first tool in the right hand column is used to drag out rectangular areas. The second tool in the right hand column is used to define oval or circular areas. Using the third tool in the right hand column, you can connect a sequence of points with line segments, creating a polygon of your own design. To close the polygon (connect your last point to your first point), just click the mouse twice in rapid succession. The fourth selection in the right hand column allows you to freehand a curve.

Approximate the area of the caldera using each selection tool as accurately as possible. After making each selection, first Measure then Show Results from the Analyze pull-down menu. Record your measurements

- using the rectangle tool
- using the oval tool
- using the polygon tool
- using the freehand tool (not the pencil in the left hand column)

Which measurement do you think is most accurate? Why?

Approximate the area of the entire volcano using each selection tool. After making each selection, first Measure then Show Results from the Analyze pull-down menu. Record your measurements

- using the rectangle tool
- using the oval tool
- using the polygon tool
- using the freehand tool (not the pencil in the left hand column)

Which measurement do you think is most accurate? Why? Which is bigger, Olympus Mons or the state in which you live?

Summary

Projecting a 3-dimensional object, such as a planetary surface, onto a 2-dimensional display, such as a computer screen, inevitably leads to distortions and loss of data. These distortions make certain types of measurement difficult if not meaningless. Image processing programs such as NIH Image assume that all pixels have the same size foot print. While this assumption is generally not true, the magnitude of the errors introduced varies from one situation to another depending on a number of factors, including the position of the observer relative to the planetary surface and the portion of the surface under study. These and other factors make the geometry of remote sensing a challenging and interesting topic.

URLs		Note: Begin each URL with the prefix http://
6.6.1	c	rst.gsfc.nasa.gov/TofC/Coverpage.html
6.6.2	c	www.math.montana.edu/~nmp/materials/ess/rs/index.html
6.6.3	h	observe.ivv.nasa.gov/nasa/education/teach_guide/ remote_history.html
6.6.4	d	sedac.ciesin.org/entri/rsimages/RSGallery.html
6.6.5	d	observe.ivv.nasa.gov/nasa/education/tools/ sources_data_earth.html
6.6.6	c	www.cipe.com/
6.6.7	d	www.msss.com/newhome.html
6.6.8	s	www.scioncorp.com/frames/fr_download_now. htm

Table 6.6.2 Section 6.6 URLs (c = concept, h = history, s = software, d = data)

Exercises

This section should be treated as a guided investigation. Perform all the actions indicated and record all measurements for discussion in class.

6.7 Applications of Geometry in Terrain Rendering

In this section, you will . . .

- **Learn the mathematical basis for estimating elevation;**
- **Investigate uses of DEM data using the image processing program** *Scion Image;*
- **Investigate uses of DEM data using terrain rendering software.**

For much of this century, aerial photography has provided detailed information on both natural and man-made features of the landscape. In recent years, satellite-based photography has been used to create a comprehensive photographic record of the Earth's surface. Landscape features of interest to scientists, engineers, and other professionals include topography, vegetation, water in its various forms, mineral composition, and man-made structures. For instance, aerial photography has been used extensively this century to estimate the height of objects on the ground. This section begins by discussing two methods for making such estimates.

Measuring Height from Above

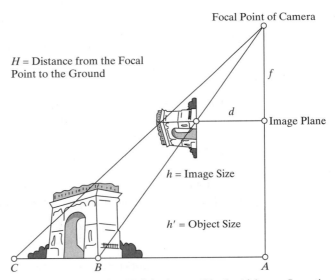

Figure 6.7.1 Measuring Height from a Single Airborne Location

Figure 6.7.1 illustrates the relationship between an object on the ground and its image as seen from the perspective of an airborne camera. All lines are assumed to lie in the same plane. Because the camera is pointed straight down, the point directly below the camera (A) is the center of every photograph. Photographs are taken as objects on the ground enter the camera's field of view. Using similar triangles, a formula may be developed that relates the height of an object on the ground to the size and position of its image on the film. The following elements of Figure 6.7.1 are used in developing the formula:

H Vertical distance from the focal point of the camera to the ground

f Vertical distance from the focal point of the camera to the film

d Film distance from the center of the image to the base of the object

h Film distance from the base to the top of the object

h' Actual distance from the base to the top of the object

AB Ground distance from a point directly below the plane to the object

AC Ground distance from a point directly below the plane to a point directly behind the top of the object as seen from the perspective of the observer

Using properties of similar triangles, we may write

$$\frac{f}{H} = \frac{d}{AB}; \quad \frac{f}{H} = \frac{d+h}{AC}; \quad \frac{d}{AB} = \frac{d+h}{AC}; \quad \frac{d}{d+h} = \frac{AB}{AC}; \quad \frac{h'}{BC} = \frac{H}{AC}$$

Using properties of proportions and rearranging terms, we obtain

$$\frac{h'}{H} = \frac{BC}{AC} = \frac{AC-AB}{AC} = 1 - \frac{AB}{AC} \Rightarrow \frac{AB}{AC} = 1 - \frac{h'}{H}$$

Substituting and further rearrangement yields

$$\frac{d}{d+h} = 1 - \frac{h'}{H} \Rightarrow \frac{h'}{H} = 1 - \frac{d}{d+h} = \frac{h}{d+h} \Rightarrow \frac{h'}{H} = \frac{h}{d+h} \Rightarrow h' = H\frac{h}{d+h}$$

It is significant that no ground measurements or camera specifications appear in the final formula. The only data required other than measurements (d and h) taken directly from the photographs is the altitude of the camera, H.

Example 6.7.1   Assuming that $H = 300$ m, $h = 2$ cm, and $d = 4$ cm, the actual height of the object in the photograph may be determined as follows:

$$(300)\left(\frac{2}{2+4}\right) = 100 \text{ m}$$

Figure 6.7.2 illustrates a related task, determining the height, h, of an object on the ground using measurements taken from two airborne positions, A and B. All lines are assumed to lie in the same plane.

d Distance between A and B

h Height of the object on the ground

τ_1 Angle subtended by the object as seen from Position 1

τ_2 Angle subtended by the top of the object and Position 2

ρ_1 Angle subtended by the object as seen from Position 1

ρ_2 Angle subtended by the top of the object and Position 1

δ_1 Angle subtended by AC as seen from the top of the object

δ_2 Angle subtended by BC as seen from the top of the object
α Angle subtended by AB as seen from the top of the object
β Angle deviation of Position 2 from the vertical as seen from the base of the object
γ Angle deviation of Position 1 from the vertical as seen from the base of the object

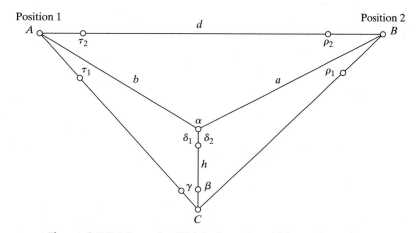

Figure 6.7.2 Measuring Height from Two Airborne Locations

The following angle relationships are based on the angle sums of various triangles:

$$\sigma = \tau_1 + \tau_2 + \rho_1 + \rho_2; \gamma = \pi - \sigma - \beta; \alpha = \pi - \rho_2 - \tau_2$$

Using the Law of Sines, we may write

$$\frac{\sin \rho_2}{b} = \frac{\sin \tau_2}{a}; \frac{\sin \gamma}{b} = \frac{\sin \tau_1}{h}; \frac{\sin \beta}{a} = \frac{\sin \tau_1}{h}; \frac{\sin \alpha}{d} = \frac{\sin \tau_2}{a}$$

Substitutions and combinations of terms yield the expressions

$$h = a\frac{\sin \tau_1 \sin \rho_2}{\sin \gamma \sin \tau_2} \quad \text{and} \quad a = d\frac{\sin \tau_2}{\sin \alpha}$$

Further rearrangements lead to the expressions

$$h = -d\left[\frac{\sin \tau_1 \sin \rho_2}{\sin(\rho_2 + \tau_2)\sin \gamma}\right] \quad \text{where } \gamma = \cot^{-1}\left[-\cot \sigma - \frac{1}{\sin \sigma}\left(\frac{\sin \tau_2 \sin \rho_1}{\sin \rho_2 \sin \tau_1}\right)\right]$$

The success of this approach depends on precise knowledge of the value of d, the distance between Positions 1 and 2. In modern satellite systems, this value is often known within a matter of centimeters. Angle measurements are also known with great precision. When flying over the surface of a planet other than the Earth, determining the satellite's altitude can be problematic. This approach avoids that problem.

Example 6.7.2

Assume that $d = 13{,}600$ m, $\tau_1 = 9°$, $\tau_2 = 29°$, $\rho_1 = 9°$, and $\rho_2 = 26°$, and the computed value of h is 1,520 m.

We now turn to the use of elevation data and a popular image processing tool.

Investigating the Topography of Mt. Adams

Topographical maps provide detailed information about landscape features not normally indicated on conventional maps. Used extensively by hikers and other backcountry travelers, "topo" maps typically use contour lines and/or different shadings to indicate different elevations.

Figure 6.7.3 is a topo map of Mt. Adams in the State of Washington, an area approximately 7680 m on a side. The image is constructed from a data array with 256 rows and 256 columns. Each element in the data array is an elevation measured in meters. The range of elevations in the data set is divided into 13 subsets, each with its own shading. The lowest elevation, 1848 m, corresponds to the dark shading found in the corners of the image. Adjacent shadings differ in elevation by 150 m. Estimate the elevation of the white region at the center of the image. How does the use of only 13 shadings limit the information presented in the map?

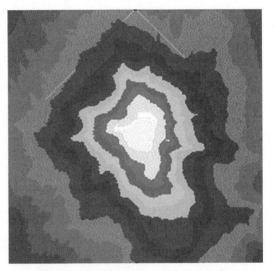

Figure 6.7.3 Mt. Adams Topography

Figure 6.7.4 is also based on a 256 × 256 array. In this case, the range of elevations is divided into 256 subsets, each with its own shading. By increasing the number of shadings from 13 to 256, a more detailed view of the terrain is obtained. When examined with an image processing program, such as *Scion Image,* this digital elevation model (DEM) may be used to answer a variety of questions about the terrain around Mt. Adams. What does the mountain look like from different perspectives? Where is the highest point? How does the elevation change along a line? How much of the mountain is above 3000 m?

CD
6.7.1

Start *Scion Image.* Select **Open** in the **File** pull-down menu. When the window appears, select Adams.tif and click **Open.** You should see a

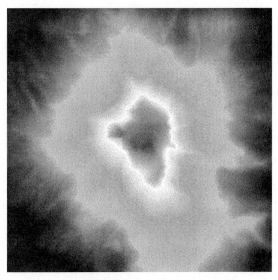

Figure 6.7.4 Mt. Adams DEM Model

computer model similar Figure 6.7.4, except in color. Compare the color image on your computer with Figure 6.7.3. Which image do you find more interesting? Which appears to provide greater detail? Which do you find more helpful in imagining the actual shape of the mountain?

A perspective view (see Figure 6.7.5) of the data is obtained by selecting the **Surface Plot** option in the **Analyze** pull-down menu. Other views may be created by clicking on the Adams.tif image, selecting one of the **Rotate** options in the **Edit** pull-down menu, and repeating the **Surface Plot** procedure. What additional insights into the shape or form of the mountain does this view provide? Print out two different perspective views of Mt. Adams and identify features that are visible in both views, and features that are visible in one view but not the other.

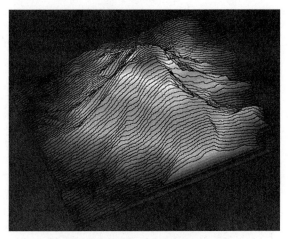

Figure 6.7.5 Mt. Adams Perspective

Before using *Scion Image* to measure length or separation, you must perform several actions. First, select the **Set Scale** option in the **Analyze** pull-down menu. When the window opens, fill in the **Measured Distance, Known Distance,** and **Units** as shown in Figure 6.7.6. [Recall that the image width of 256 pixels corresponds to a distance of 7680 m in the real world.]

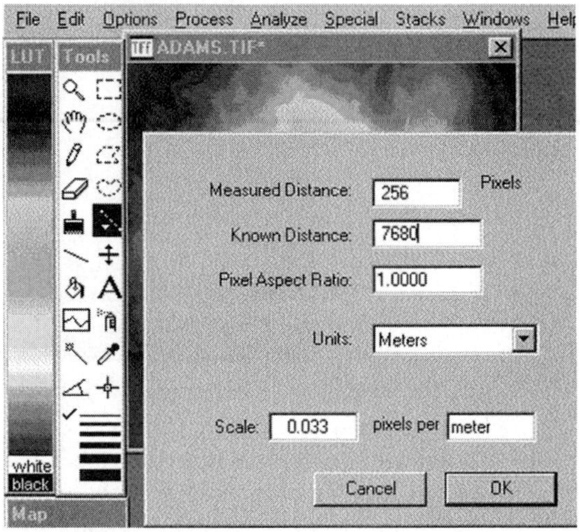

Figure 6.7.6 Set Scale

Second, select **Options** in the **Analyze** pull-down menu. When the window opens, use the mouse to check the **Perimeter/Length** box, as shown in Figure 6.7.7. Remove any other checks.

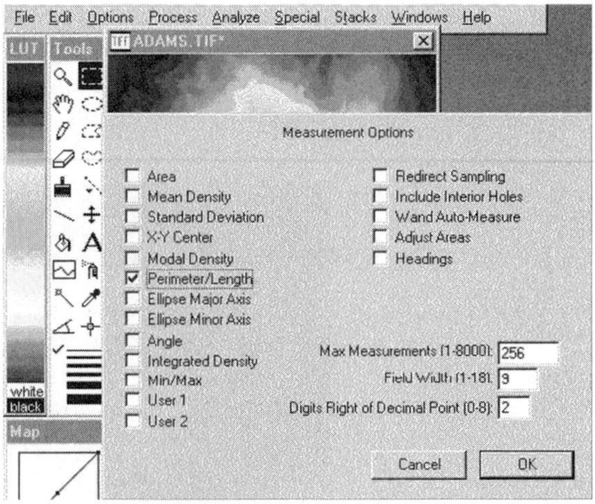

Figure 6.7.7 Measurement Options

Third, select the line tool (darkened in the **Tools** menu in Figure 6.7.8) and use it to draw a line between two points. In the **Analyze** pull-down menu, select **Measure** then **Show Results.** The measurement appears in a **Results** window. Take a series of measurements using this procedure: The

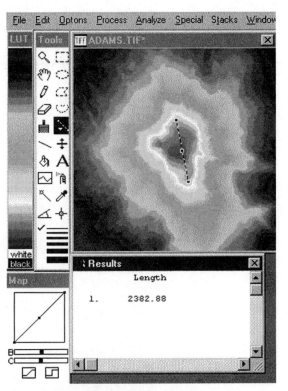

Figure 6.7.8 Measuring Distance

length and width of the image; the diagonal of the image; the greatest distance between pixels shaded as shown.

A similar procedure is used to measure perimeter and area. Select **Options** in the **Analyze** pull-down menu. When the window opens, check the **Perimeter/Length** box. Next, select the freehand drawing tool (darkened in the **Tools** window in Figure 6.7.9) and use it to sketch a closed curve. Finally, use the **Measure** and **Show Results** procedures as before. Use this procedure to find the perimeter and area of other regions.

Data in this model are integers ranging in value from 2–242. Different data values are associated with different shadings in the **LUT,** or look up table. Select the cross-hair tool in the **Tools** window and run it over the **LUT.** The data value associated with each shading appears in the **Info** window as an **Index.** These data values represent elevations ranging from 1848 m–3753 m. Use this information to develop a formula to convert data values (2–242) to elevations (1848–3753). When a line is drawn in the model with the plot profile tool (darkened in the **Tools** window, Figure 6.7.10), a curve is generated showing the data values along that line. The curve may be thought of as the elevations a hiker would pass through in traversing the mountain along the indicated path. Plot a series of profiles based on vertical lines drawn at even intervals across the image. Use the formula you developed to find the highest elevation along each path. Which profile appears to be the longest?

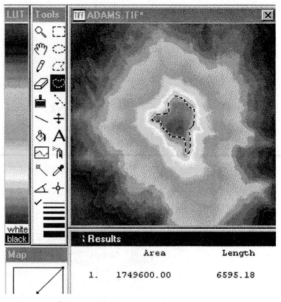

Figure 6.7.9 Area and Perimeter

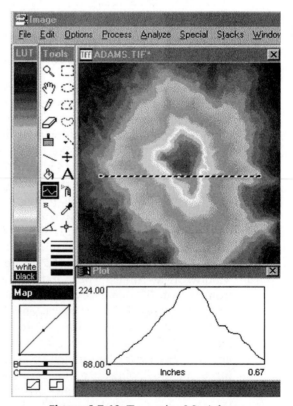

Figure 6.7.10 Traversing Mt. Adams

Select **Threshold** in the **Options** pull-down menu and the LUT tool (see Figure 6.7.11, darkened in the **Tools** window). Move the **LUT** tool into the LUT window. Hold down the mouse key as you move the LUT tool up and down. As you do so, watch the **Index** values in the **Info** window. Thresholding shows all values in the model at or above the current **Index** value. Use the elevation formula you developed to compute the Index values associated with the following elevations: 2000 m; 2500 m; 3000 m; 3500 m. Position the LUT tool at each of these values and find the area of each black region. Select the region using the wand tool in the **Tools** window, then use the **Measure** and **Show Results** options in the **Analyze** pull-down menu.

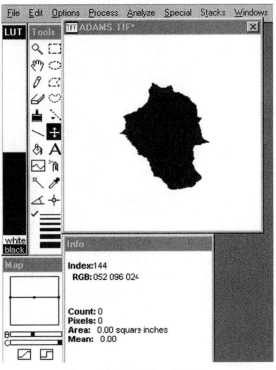

Figure 6.7.11 Thresholding

A histogram of the Mt. Adams (see Figure 6.7.12) data shows the relative frequency of elevations in the model. Select **Load Macros** in the **Special** pull-down menu. When the window opens, select **Plotting Macros.** Return to the **Special** pull-down menu and select **Histogram.** The result is displayed in a **Histogram** window. What is the most frequent data value in the model? What elevation does this data value represent?

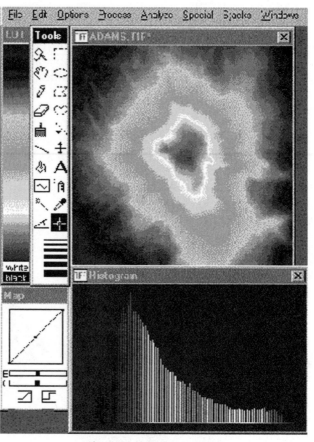

Figure 6.7.12 Histogram

While *Scion Image* provides a number of powerful measurement options, it is fundamentally an image processing tool, not a terrain rendering tool. The following example illustrates the use of tools developed by the US Geological Survey for viewing digital elevation data.

Modeling Mountainous Terrain

URL 6.7.1

The US Geological Survey (USGS) maintains a database of US topographical data. Some of these data are available for free. For instance, a database of 1:250,000-scale Digital Elevation Models (DEM) is available on-line. These DEM models are available for every state in both compressed and uncompressed formats. Because the uncompressed versions are often 10 MB or more in size, most users download the compressed versions of the files. Using a dial-up modem, transferring the compressed files takes approximately 15 minutes. Once downloaded, the compressed files should be uncompressed using WinZip and the suffix.dem added to the resulting file's name.

A number of software developers distribute free software for viewing DEM and other cartographic data files. The following overviews of these tools provide a brief orientation to their capabilities. For students just beginning an investigation of 3-dimensional surfaces, these tools provide a motivating and intuitive basis for thinking about 3-dimensional data.

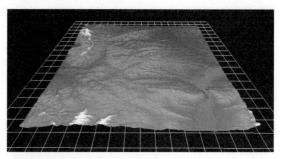

Figure 6.7.13 dem3D Rendering, Bozeman MT

**URL
6.7.2** The program *dem3D* (see Figure 6.7.13) was formally cancelled by USGS before completion. Whatever their reasons for doing so, *dem3D* offers students a limited but easy-to-use tool for displaying and modifying perspective views of DEM landscape files. The viewer may move about relative to the landscape, exaggerate the vertical scale, add compass headings, vary the angle of the lighting, and color the scene to emphasize a variety of topographic features, including elevation and gradient.

Figure 6.7.14 3dem60 Rendering, Bozeman MT

**URL
6.7.3** The program *3dem60* (see Figure 6.7.14), developed by R.S. Home, is a much more sophisticated tool than *dem3D*. Users have many more options in specifying the point of view of the observer and the coloring used, including the option of creating animated flyovers of the terrain. The URL listed in Table 6.7.1 offers extensive technical information and sample renderings.

**URL
6.7.4** The USGS developed program *dlgv32* (see Figure 6.7.15) displays a number of cartographic file formats, including DEM. While it does not provide perspective views, it does support distance measurement. The URL listed in Table 6.7.1 offers extensive technical information and a tutorial.

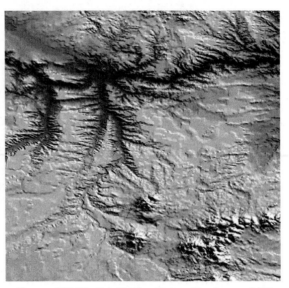

Figure 6.7.15 dlgv32 Rendering, Lewistown MT

Summary

Digital elevation models (DEM) are used extensively to study the terrain of both the Earth and the other planets of the solar system. Image processing and terrain modeling tools such as *Scion Image* and 3dem60 support a wide variety of measurement and rendering features. By using these and other tools, detailed information and accurate insights may be formulated concerning the shape and structural features of objects in the DEM.

Height of an Object in an Aerial Photograph

Q: Observed from a single location a given distance H above the ground, how may the height of an object be determined on the basis of its appearance in a photograph?

A: $h' = H \dfrac{h}{d + h}$

where H = actual distance from the camera to the ground, d = distance on film from the center of the image to the base the object, h = distance on film from the base to the top of the object.

Q: Observed from two locations a given distance apart, how may the height of an object be determined on the basis of its appearance in a photograph?

A: $h = -d \left[\dfrac{\sin \tau_1 \sin \rho_2}{\sin(\rho_2 + \tau_2)\sin \gamma} \right]$ and

$\gamma = \cot^{-1} \left[-\cot \sigma - \dfrac{1}{\sin \sigma} \left(\dfrac{\sin \tau_2 \sin \rho_1}{\sin \rho_2 \sin \tau_1} \right) \right]$ and

    d Distance between *A* and *B*
  ·*h* Height of the object on the ground
  τ_1 Angle subtended by the object as seen from Position 1
  τ_2 Angle subtended by the top of the object and Position 2
  ρ_1 Angle subtended by the object as seen from Position 1
  ρ_2 Angle subtended by the top of the object and Position 1

URLs		Note: Begin each URL with the prefix http://
6.7.1	d	edcwww.cr.usgs.gov/glis/hyper/guide/1_dgr_demfig/states.html
6.7.2	s	mcmcweb.er.usgs.gov/viewers/dem3d/
6.7.3	s	www.monumental.com/rshorne/3dem.html
6.7.4	s	mcmcweb.er.usgs.gov/viewers/

Table 6.7.1 Section 6.7 URLs (c = concept, h = history, s = software, d = data)

Exercises

1. In Figure 6.7.1, let $H = 500$ m, $h = 3$ cm, and $d = 4$ cm. Find the actual height of the object in the photograph.
2. In Figure 6.7.2, let $d = 13000$ m, $\tau_1 = 8°$, $\tau_2 = 25°$, $\rho_1 = 7°$, $\rho_2 = 27°$. Find the computed value of h.
3. The section on topography should be treated as a guided investigation. Perform all the actions indicated and record all measurements for discussion in class.
4. Download the USGS DEM file for where you live and view it using the USGS program 3dem60.

References and Suggested Reading

1:250,000-scale Digital Elevation Models (DEM). EROS Data Center, Sioux Falls, SD. USGS. Available on-line http://edcwww.cr.usgs.gov/glis/hyper/guide/1_dgr_demfig/states.html

Brock, R. and H. Ryser 1949. "The non-existence of certain finite projective planes." Canadian Journal of Mathematics, 1, 88–93.

Center for image processing in education. Available on-line http://www. cipe.com/

DesignWorkshop Lite. Artifice, Inc. Available on-line http://www.artifice.com/dw_lite.html

dlgv32: Software for viewing USGS digital cartographic maps. USGS. Available on-line http://mcmcweb.er.usgs.gov/viewers/

ENTRI Remote Sensing Image Gallery. Environmental Treaties and Resource Indicators (ENTRI). Available on-line http://sedac.ciesin. org/entri/rsimages/ RSGallery.html

Flavin, P. 1997. *Java 3-D Viewer*. Available on-line www.frontiernet.net/~imaging/java-3d-engine.html

Geomantics Ltd. *Genesis II*. Available on-line www.geomantics.com/

Horne, R. *Microcomputer Topography*. Available on-line http://www. monumental.com/rshorne/3dem.html

Mars Global Surveyor. Malin Space Science Systems, Inc. Available on-line http://www.msss.com/newhome.html

O'Connor, J.J. and E.F. Robertson. 1999. *MacTutor History of Mathematics archive.* School of Mathematics and Statistics, University of St. Andrews, Scotland. Available on-line
Alberti. Available on-line http://www-history.mcs.st-andrews.ac.uk/history/Mathematicians/Alberti.html

Leon Battista Alberti. *The Florence art guide.* Firenze by Net. Available on-line http://www.mega.it/eng/egui/hogui.htm

Pinch, R. *Coding theory: The first 50 years.* Available on-line http://pass.maths. org.uk/issue3/codes/index.html

Remote sensing in history. NASA Observatorium teacher's guide. Available on-line http://observe.ivv.nasa.gov/nasa/education/teach_guide/remote_history.html

Scion Image. Scion Corporation. Available on-line http://www.scioncorp.com/frames/fr_download_now.htm

Short, N.M. 1999. *The remote sensing tutorial.* NASA/Goddard Space Flight Center. Available on-line http://rst.gsfc.nasa.gov/Front/foreword.html

Software for viewing DEM data. USGS. Available on-line http://mcmcweb.er.usgs. gov/viewers/dem3d/

Sources of Earth remote sensing data. NASA Observatorium. Available on-line http://observe.ivv.nasa.gov/nasa/education/tools/sources_data_earth.html

Thomas, D. 1999. *Maps, models and remote sensing: From Montana to Moscow.* Network Montana Project. Montana State University. Available on-line www.math.montana.edu/~nmp/materials/ess/rs/index.html

Thomas, D., C. Thomas, and B. Beaudrie, 1999. "On-line science activities from the Network Montana Project and the NASA CERES Project." Proceedings of the International Conference on Mathematics/Science Education & Technology. San Antonio, TX, 1–4 March 1999, 549–555.

Thomas, D. and C. Thomas 1997. "Using Internet-based K–12 classroom activities: Materials & staff development." INET'97: Proceedings of the 1997 meeting of the Internet Society. Kuala Lumpur, Malaysia, 24–27 June, 1997.

Thomas, D. 1995. Getting started with scientific CD-ROMs. Educational Technology Review, 4, 7-12.

Thomas, D. (ed.) 1995. *Scientific visualization in mathematics and science teaching.* Charlottesville: Association for the Advancement of Computing in Education.

Thomas, D. and N. Sabelli 1993. "Visualization in mathematics and science teaching. Rethinking the roles of technology in education." Proceedings of the tenth international conference on technology in education, (166–168). Cambridge, MA, March 21–24, 1993.

Thomas, D. 1992. "Using computer visualization to motivate and support mathematical dialogues." Journal of computers in mathematics and science teaching, 11(3/4), 265–274.

Veblen, O. and Bossey, W. 1906. "Finite projective planes." Transactions of the American Mathematical Society, 7, 241-259.

Weber, S. 1998. *Polyhedra of Point Groups.* Available on-line http://www.nirim.go.jp/~weber/JAVA/jpoly/jpoly.html

Wolff, R. and L. Yaeger 1993. *Visualization of natural phenomena.* Santa Clara: Springer-Verlag.

Zimmerman, W. and S. Cunningham (eds.) 1991. *Visualization in teaching and learning mathematics.* Washington: Mathematical Association of America.

Index